STANDARD METHODS OF
CLINICAL CHEMISTRY

VOLUME 4

CONTRIBUTORS TO THIS VOLUME

Joseph S. Amenta · George N. Bowers, Jr. · Wendell T. Caraway · Robert L. Dryer · Frank W. Fales · O. P. Foss · Elizabeth G. Frame · Edythe R. Gershman · Thomas J. Giovanniello · Richard J. Henry · R. Hobkirk · George W. Johnston · Bernard Klein · Ann Metcalfe-Gibson · Theodore Peters, Jr. · Ralph E. Peterson · John G. Reinhold · Eugene W. Rice · J. I. Routh · Robert H. Silber · Warren M. Sperry · Rex E. Sterling · Clifford B. Walberg · Arnold G. Ware · Herbert Weissbach · Howard J. Wetstone · Virginia L. Yonan

STANDARD METHODS OF
CLINICAL
CHEMISTRY

VOLUME 4

By the American Association of Clinical Chemists

Editor-in-Chief

DAVID SELIGSON

Director of Clinical Laboratories
Grace-New Haven Community Hospital
Yale-New Haven Medical Center
Associate Professor of Medicine
Yale University School of Medicine

1963

ACADEMIC PRESS • New York and London

ACADEMIC PRESS INC.
111 Fifth Avenue
New York 3, New York

United Kingdom Edition
Published by
ACADEMIC PRESS INC. (London) Ltd.
Berkeley Square House, London W. 1

Library of Congress Catalog Card Number: 53-7099

This volume is dedicated to John Reinhold (1900-   ), a teacher
and colleague of the editor, who has inspired many with his teaching
and scientific contributions and who, with his kind, generous, gentle-
manly and scholarly ways, is now guiding students in the Near East.

CONTRIBUTORS

JOSEPH S. AMENTA, *Yale University, New Haven, Connecticut and Walter Reed Army Medical Center, Washington, D.C. (31)*

GEORGE N. BOWERS, JR., *Departments of Pathology and Medicine, Hartford Hospital, Hartford, Connecticut (47, 163)*

WENDELL T. CARAWAY, *Flint Medical Laboratory, and Laboratories of McLaren General Hospital and St. Joseph Hospital, Flint, Michigan (239)*

ROBERT L. DRYER, *Department of Biochemistry, State University of Iowa, Iowa City, Iowa (191, 205)*

FRANK W. FALES, *Department of Biochemistry and Clinical Research Center, Emory University, Atlanta, Georgia (101)*

O. P. FOSS,* *Medical Research Center, Brookhaven National Laboratory, Upton, New York (125)*

ELIZABETH G. FRAME, *Clinical Pathology Department, Clinical Center, National Institutes of Health, Bethesda, Maryland (1)*

EDYTHE R. GERSHMAN, *William Pepper Laboratory of Clinical Medicine, Hospital of the University of Pennsylvania, Philadelphia, Pennsylvania (85)*

THOMAS J. GIOVANNIELLO, *Laboratory Service, Boston Veterans Administration Hospital, Boston, Massachusetts (139)*

RICHARD J. HENRY, *Director, Bio-Science Laboratories, Los Angeles, California (205)*

R. HOBKIRK, *McGill University Medical Clinic and Department of Metabolism, The Montreal General Hospital, Montreal, Canada (65)*

GEORGE W. JOHNSTON, *Third U.S. Army Medical Laboratory, Fort McPherson, Georgia (183)*

BERNARD KLEIN, *Veterans Administration Hospital, Bronx, New York (23)*

ANN METCALFE-GIBSON, *McGill University Medical Clinic and Department of Metabolism, The Montreal General Hospital, Montreal, Canada (65)*

---

* Present address: Norwegian Radium-hospital, Oslo, Norway.

viii CONTRIBUTORS

THEODORE PETERS, JR., *Mary Imogene Bassett Hospital, Cooperstown, New York (139)*

RALPH E. PETERSON, *Cornell Medical College, New York, New York (151)*

JOHN G. REINHOLD, *William Pepper Laboratory of Clinical Medicine, Hospital of the University of Pennsylvania, Philadelphia, Pennsylvania (85)*

EUGENE W. RICE, *W. H. Singer Memorial Research Laboratory of the Allegheny General Hospital, Pittsburgh, Pennsylvania (39, 57)*

J. I. ROUTH, *Department of Biochemistry, University of Iowa School of Medicine, Iowa City, Iowa (191)*

ROBERT H. SILBER, *Merck Institute for Therapeutic Research, Rahway, New Jersey (113)*

WARREN M. SPERRY, *Departments of Biochemistry, New York State Psychiatric Institute and the College of Physicians and Surgeons, Columbia University, New York, New York (173)*

REX E. STERLING, *Chemistry Division of the Main Laboratory, Los Angeles County Hospital, Los Angeles, California, and Department of Biochemistry and Nutrition, University of Southern California, School of Medicine, Los Angeles, California (15)*

CLIFFORD B. WALBERG, *Chemistry Division of the Main Laboratory, Los Angeles County Hospital, Los Angeles, California, and Department of Biochemistry and Nutrition, University of Southern California, School of Medicine, Los Angeles, California (15)*

ARNOLD G. WARE, *Chemistry Division of the Main Laboratory, Los Angeles County Hospital, Los Angeles, California, and Department of Biochemistry and Nutrition, University of Southern California, School of Medicine, Los Angeles, California (15)*

HERBERT WEISSBACH, *Laboratory of Clinical Biochemistry, National Heart Institute, National Institutes of Health, Public Health Service, U.S. Department of Health, Education, and Welfare, Bethesda, Maryland (121, 197)*

HOWARD J. WETSTONE, *Departments of Medicine and Pathology, Hartford Hospital, Hartford, Connecticut (47)*

VIRGINIA L. YONAN, *William Pepper Laboratory of Clinical Medicine, Hospital of the University of Pennsylvania, Philadelphia, Pennsylvania (85)*

FOREWORD TO THE SERIES

This series of "Standard Methods of Clinical Chemistry" was successfully started with the publication of Volume I in 1953 under the editorship of Dr. Miriam Reiner. The policies established, as listed in the Preface and Foreword of Volume 1, have been continued.

The clinical chemist usually works in a hospital milieu. His interests are service, research, and teaching. When these are conscientiously pursued, the quality of medical care and benefits to the medical staff and patients improve. The range of knowledge and skills required in the practice of clinical chemistry is wide, and often the busy or less experienced analyst does not have the time to investigate new or difficult methods which he desires to use. This series is designed to provide accurate and workable methods upon which the clinical chemist can rely. Although some of these methods may not be used for daily work they should be helpful as references for evaluating those in use. This series is directed primarily to clinical chemists. However, we believe that pathologists, medical technicians, clinical investigators, chemists in other fields, students, and others will find in this volume, useful methods, ideas, facts, and references.

In the choice of methods, accuracy is a guiding principle. Some are very easy to perform. Others are time-consuming but essential for the practice of medicine. Some are more difficult than methods currently in use but have the advantage of being accurate and useful as reference methods.

Several methods included in the series are acknowledged to be more time consuming than some in common use. However, they were chosen because they provide dependable performance. Several of our methods offer alternatives or short cuts with adequate warnings. References to methods which the analyst might find useful because of his special problems are included as often as possible.

The quality of reagents used, purification, and primary and secondary standards have been described. When doubt exists reagent grade chemicals should be used. "Water" refers to distilled water. We have tried to use significant figures to indicate the range of accuracy required for reagents or measurements. Where information was available, stability of reagents is stated. Information on collection of specimens and storage is provided, wherever it is essential or available. Notes have been freely interspersed in the text to allow the contributor or referee to make clear his directions, specifications or point of view. References made to commercial sources of equipment or materials do not imply that competitive products were less adequate.

Discussion of clinical or physiological problems related to the subject of the various chapters has been kept to a minimum. Often pertinent references to clinical and physiological reviews have been included. Ranges of values in healthy persons and often in disease, as established by these methods, have been included. We hope that analysts will report their data to us so that we can accumulate these for statistical studies.

To those clinical chemists who study this series, contribute to its improvement, and regard these methods as stepping stones to better ones, we offer our thanks.

DAVID SELIGSON, Editor

PREFACE

Modern instruments are getting better all the time. Better refers to increased accuracy, reliability and ease of use. Methods in this volume involve the use of spectrofluorimeters, ultraviolet, visible and infrared spectrophotometers. Good instruments are tools of the analyst and not vice versa. The analyst uses these instruments and others to measure a property of matter. Sometimes, this property has such specificity that the measurement may be made in crude samples. More often, however, it is necessary to put them through one or more isolation steps such as a protein separation, paper chromatographic separation, distillation and so on to increase the specificity of the procedure. Sufficient purification provides material which can be measured by a non-specific property such as weight. In this category is the measurement of chloride as silver chloride.

The skilled analyst knows the properties of matter and his chemical reactions sufficiently well to make certain assumptions which reduce his technical operations while providing adequately accurate analyses. Scientific knowledge and experience establish the background for the skilled chemist so that his assumptions and concessions are sound. He also knows how to prove these when necessary. The technician who lacks this fundamental knowledge is a recipe-follower and a knob-turner. Not infrequently, business men enter a laboratory and give advice on the purchase of instruments and ready-made reagents and the use of methods. In this instance, the manufacturer is making the assumptions and concessions in instrumentation and methodology, not the analyst. The indifferent analyst can buy the reagents and protocols for the measurement of many constituents in serum in "kit" form. He may not even know the original author of the procedure or the manufacturer's secret ingredients. In some situations, the manufacturer or the sales representative may have misled him by clever advertising and sales devices. The latter individual, with little scientific knowledge and much superficial sophistication, cares more for his commission than the accuracy of his knowledge. The current success of many firms selling kits, reagents, methods and special instruments attests to the large number of customers. A reprehensible and not uncommon practice is the publications of research work in scientific journals which uses undocumented commercial methods.

It is hoped that this volume will provide the reader with suitable details for performing the analyses described and references to the literature which may further his understanding of the chemical reactions involved or provide alternative methods. Where possible, the chapters or their lists of references contain the proof of the methods. We hope these may serve as references for comparison with the advertised kits, automated methods and short-cut procedures. Clinical Chemists have a serious obligation to strive for improvement of these analytical procedures in order to achieve accuracy in methodology (errors of less than 1.0 per cent) which can be used for reference purposes.

DAVID SELIGSON, *Editor*

New Haven, Connecticut
October, 1963   *Editorial Committee*
                                         MARGARET KASER
                                         MIRIAM REINER
                                         JOHN REINHOLD
                                         JOSEPH ROUTH

CONTENTS

xiii

xiv CONTENTS

STANDARD METHODS OF
CLINICAL CHEMISTRY

FREE AMINO ACIDS IN PLASMA AND URINE BY THE GASOMETRIC NINHYDRIN-CARBON DIOXIDE METHOD*

Submitted by: ELIZABETH G. FRAME, Clinical Pathology Department, Clinical Center, National Institutes of Health, Bethesda, Maryland
Checked by: NEIL Y. CHIAMORI, Bioscience Laboratories, Los Angeles, California
PAULINE HALD and ETHEL CONGER, Grace-New Haven Community Hospital, New Haven, Connecticut

Introduction

Although simpler and more rapid methods for the determination of amino acids in plasma and/or urine have been described (e.g., 1, 2), none has the specificity of the gasometric ninhydrin method described by Van Slyke and his collaborators (3, 4, 5). This reaction is specific for free amino acids in that it requires the presence, in the free unconjugated state, of both the carboxyl and the neighboring NH_2 or $NH–CH_2$ group. All of the common amino acids, including proline and hydroxyproline, yield 1 mole of CO_2 per mole of α-amino nitrogen except for aspartic acid which yields 2 moles. Proteins, peptides (other than glutathione, which has a free α-amino group adjoining a free carboxyl group), and most substances other than amino acids do not react significantly. Urea reacts to a small extent and must be corrected for, or removed from, plasmas with high urea concentrations and urines.

Principle

The sample, at pH 2 to 3, is heated at 100°C. in a closed reaction vessel with ninhydrin (triketohydrindene hydrate). The amino acids react with the formation of ammonia, CO_2, and aldehydes. (At a pH above 4, a blue color is formed owing to condensation of the liberated ammonia with ninhydrin and its reduction product.) The CO_2 is transferred to the chamber of the Van Slyke-Neill manometric apparatus and there measured. From the amount of CO_2 found, the α-amino nitrogen content of the sample may be calculated.

* Based on the method of Van Slyke and his collaborators (3, 4, 5).

1

METHOD

Apparatus

1. Van Slyke-Neill manometric apparatus.

2. Storage vessel for CO_2-free 0.5 N NaOH. A 250-ml. separatory funnel provided with a rubber stopper holding a soda-lime tube is satisfactory. The tip of the funnel dips beneath the surface of mercury in a 50-ml. Erlenmeyer flask, the mercury being covered with a few milliliters of 1 N H_2SO_4.

3. Reaction vessels, all glass, Van Slyke type. These are available from a number of supply houses. Checker N.Y.C. used reaction vessels 18 mm. I.D. × 100 mm. fused to 6 mm. I.D. × 50 mm. I.D. tubing. The latter was connected to rubber tubing 3/16-inch wall, 3/16-inch bore and 6 cm. long. The vessels were sealed off during the reaction with a screw clamp and glass plug (8 mm. × 25 mm.).

4. Stopcock pipette, to deliver 1 ml., and provided with a rubber tip.

5. Glass spoon, to deliver 100 ± 10 mg. ninhydrin.

6. Rubber connecting tubes. Cut 5-cm. lengths of rubber tubing (⅛-inch wall, ⅛-inch bore) and, before use, remove the CO_2 from the pores as follows. Boil the tubes in acidified water in a round-bottomed flask for 30 minutes. Immediately stopper the flask and cool under the tap. The vacuum formed in the flask draws the gases from the rubber. After the bubbles stop coming out of the rubber, open the flask and wash the tubes with distilled water. Repetition of the treatment for the life of the tubing is not necessary.

Reagents

1. *Picric acid, 1% solution.* Dissolve 10.0 g. of picric acid (containing 10–10.5% water), reagent grade, in water and dilute to 1 l. with water.

2. *NaOH, saturated solution (about 18 N).* Dissolve solid NaOH in an equal weight of water, and allow the solution to stand until the carbonate settles. Standardize the solution.

3. *NaOH, approximately 0.5 N in 25% NaCl, of minimal CO_2 content.* Boil about 300 ml. of distilled water for about 5 minutes in order to remove the CO_2, and cover the container with a watch glass. Cool to room temperature. Transfer 62.5 g. of NaCl, reagent grade, to a

250-ml. volumetric flask, and dissolve in, and dilute to volume with, the CO_2-free water.

Fill a 250-ml. volumetric flask to within about 10 ml. of the mark with the 25% NaCl solution and add the calculated amount of the concentrated NaOH solution from a graduated pipette whose tip dips beneath the surface. Fill the flask to the mark with the 25% NaCl solution. Stopper, mix thoroughly, and transfer the solution immediately to the storage vessel protected from CO_2. Checker N.Y.C. added 2 drops of 1% alizarin sulfonate at this point.

4. Lactic acid, approximately 2 N in 25% NaCl. Dilute 1 volume of concentrated lactic acid, reagent grade to 5 volumes with 25% NaCl solution.

5. NaOH approximately 5 N. Dilute the standardized concentrated NaOH solution with the appropriate amount of water.

6. Ninhydrin.

7. Octyl alcohol.

8. Grease for reaction vessels. Cello-Seal (Fisher) is satisfactory.

9. Alundum Norton 14x. Checker N.Y.C. used 3 mm. glass heads.

Additional Apparatus and Reagents for Determination in Urine

1. Glass spoons to deliver 170 ± 15 mg. of phosphate buffer and 100 ± 10 mg. of citrate buffer.

2. Urease, 1% aqueous solution. A freshly prepared solution of Urease, Sigma, Type II, is satisfactory. Checker N.Y.C. preferred a 2% urease solution made as follows: Urease 200 mg. (Sigma) was dissolved in 5 ml. water to which 5 ml. of glycerol was added.

3. Brom thymol blue (Na salt), 0.04% solution in water. Mix 0.1 g. of the dry indicator in a mortar with 14.3 ml. of 0.01 N NaOH. Dilute to 250 ml. with water.

4. Brom cresol green (Na salt), 0.04% solution in water. Mix 0.10 g. of the dry indicator in a mortar with 16.0 ml. of 0.01. N NaOH. Dilute to 250 ml. with water.

5. Solid phosphate buffer for pH 6.2; 3 parts by weight of KH_2PO_4 (anhydrous) and 1 part of Na_2HPO_4 (anhydrous). Grind the two phosphates separately and then together. Checker N.Y.C. preferred a buffer solution: 12.5 g. KH_2PO_4 and 4.0 g. Na_2HPO_4 were dissolved in 88 ml. water to a final volume 91.5 ml. One ml. is equivalent to 175 mg. of the dry buffer mixture.

6. *Solid citrate buffer for pH 2.5;* 2.06 g. of sodium citrate ($Na_3C_6H_5O_7 \cdot 2H_2O$) and 19.15 g. of citric acid ($C_6H_8O_7 \cdot H_2O$). Grind separately and then together. Checker N.Y.C. preferred a buffer solution: 2.06 g. sodium citrate and 19.15 g. citric acid to a final volume of 106 ml. One ml. is equivalent to 200 mg. of the dry buffer mixture.

7. *NaOH, 1 N.*

8. *H_2SO_4, 1 N.*

9. *H_2SO_4, 5 N.*

I. PLASMA

Procedure

PRECIPITATION OF PROTEINS

To 1 volume of plasma in a centrifuge tube add 5 volumes of 1% picric acid solution. Duplicate or triplicate analyses are recommended.

NOTE 1: Plasma rather than serum should be used since serum contains a higher concentration of amino acids than does plasma (6). Free amino acids are apparently liberated in the clotting process. Either oxalate or heparin may be used as an anticoagulant.

NOTE 2: When the urea nitrogen concentration of the plasma is higher than 20 mg. per cent, an additional step must be introduced into the procedure, as described below.

NOTE 3: Precipitation of proteins with picric acid yields a filtrate of pH 1.8 to 2.0, which is suitable for the ninhydrin reaction. Additional buffer is not required.

NOTE 4: For maximum accuracy, the proteins should be precipitated and the filtrate treated with ninhydrin promptly after the blood is drawn. The writer, however, has not found significant changes if the plasma is promptly frozen and analyzed at a later time.

Shake or stir the mixture vigorously and centrifuge for 10 minutes at 3000 r.p.m. Decant the supernatant solution through a funnel containing a plug of cotton about the size of a pea.

REMOVAL OF PREFORMED AND LABILE CO_2, ADDITION OF NINHYDRIN AND EVACUATION OF AIR

Transfer 5.0 ml. of the filtrate to a reaction vessel containing two or more alundum pieces and a drop of octyl alcohol. Bring the contents to a boil over the free flame of a micro burner in about 30 seconds, and boil for exactly 1 minute. Place the vessel in ice water for exactly 2 minutes. During this interval slip a rubber connecting tube on the side arm and wipe the ground glass surface dry. Add 100 ± 10 mg. of ninhydrin from a calibrated glass spoon and set into place the greased glass stopper, with the opening to the side arm open. As quickly as possible after the addition of the ninhydrin, evacuate the vessel to 30 mm. or less of mercury pressure. This may be accomplished with a pump or with the Van Slyke-Neill apparatus. If the latter is used, attach the vessel to the curved capillary outlet of the chamber and lower the mercury three times to the bottom of the chamber, ejecting each time the gases that are drawn over from the vessel. After evacuation is completed, close the vessel by rotating the stopper through 180°. Press the rubber tube flat as the vessel is removed from the evacuation system and insert a glass plug, trapping as little air as possible. Secure the glass stopper by anchoring it to the side arm with a rubber band.

IMMERSION IN WATER BATH AT 100°C.

Set the vessels upright in a rack or wire basket, and place in an actively boiling water bath for exactly 20 minutes with the vessels completely immersed. Remove the vessels from the bath, turn the glass stoppers 5° or 10°, and cool to room temperature. The CO_2 contents may be determined at a convenient time.

ABSORPTION OF CO_2 BY ALKALI IN VAN SLYKE-NEILL CHAMBER

Remove from the side arm of the vessel any CO_2 which may have evolved during the boiling. This is done by removing the glass plug and attaching the rubber tube to the curved capillary outlet of the Van Slyke-Neill chamber. Lower the mercury once to the bottom of the chamber and eject the gases.

Admit 2 ml. of the 0.5 N NaOH (reagent 3) into the chamber, seal the stopcock and capillary with mercury, and lower the mercury to the middle of the chamber. Open the vessel stopper as well as the stopcock at the top of the chamber so that the gas may pass from the vessel to the chamber. The CO_2 is now transferred at room temperature from the vessel to the alkali in the chamber by raising and lowering the mercury in the chamber ten times. During each lowering, shake the vessel by hand. After the last upward excursion, lower the mercury to the middle of the chamber and close the stopcock at the top of the vessel and the one leading to the leveling bulb. Remove the

reaction vessel and seal with mercury the capillary to which it has
been attached, being sure that a solid column of mercury is introduced.

EJECTION OF UNABSORBED GASES

With the stopcock at the top of the chamber closed, and with the
stopcock to the leveling bulb open, apply positive pressure to the gases
in the chamber by raising the leveling bulb a little above the top
stopcock. With the bulb at this level, close the stopcock leading to the
leveling bulb and open the top stopcock to connect the chamber with
the cup above it. Admit mercury from the leveling bulb into the
chamber until the alkali solution just reaches the bottom of the top
stopcock. Close both stopcocks. Open the lower stopcock, lower the
leveling bulb, and admit a little mercury from the cup into the chamber in order to seal the connecting capillary. The small bubble of air
trapped in the capillary is readmitted, but it is CO_2-free and is therefore without influence on the CO_2 determination.

EXTRACTION OF CO_2 AND READING OF p_1

Add 1 ml. of 2 N lactic acid (reagent 4) to the chamber, using a
stopcock pipette with a rubber tip and making the admission through
a mercury seal. Seal the capillary with mercury. Lower the mercury
in the chamber to the 50 ml. mark and close the stopcock to the
leveling bulb. Shake the chamber for 3 minutes at 300 to 400 complete excursions per minute. Readmit mercury from the leveling bulb
until the gas volume reaches 0.5 ml. The mercury should be admitted
in 30 to 40 seconds, avoiding oscillations of the mercury due to jerky
openings and closings of the stopcock. Read the pressure (p_1) on the
manometer scale. If it is necessary or desirable to reread p_1, lower
the mercury again to the 50 ml. mark and shake for 1 minute.

REABSORPTION OF CO_2 AND READING OF p_2

Open the stopcock to the leveling bulb. Add to the cup from a
pipette 0.5 ml. of 5 N NaOH and admit it to the chamber. Add 2–3 ml.
of mercury to the cup and admit most of it rather quickly to the
chamber. This serves to mix the contents of the chamber, as well as
to seal the capillary with mercury. Rinse the cup once with water.
With the stopcock to the leveling bulb open, lower the surface of the
mercury three times about a third of the way down the large bulb

of the chamber. Allow the gas volume to remain for about a minute slightly below the 0.5-ml. mark for complete drainage. Note the temperature of the water jacket of the chamber. Bring the gas volume to 0.5 ml. and read p_2.

WASHING CHAMBER AFTER ANALYSIS

Eject the aqueous mixture from the chamber into the cup above. Wash the chamber once with acidified water (5 ml, of concentrated lactic acid in 1 l. of water) and once with distilled water. A recommended washing procedure is to fill the cup with the wash solution, lower the level of mercury in the chamber halfway or more, and then admit the wash solution from the cup, without air.

BLANK ANALYSIS FOR c CORRECTION

The correction, c, due chiefly to carbonate in the 0.5 N NaOH, is the value of p_1-p_2 found in a blank analysis. It is usually not necessary to add the ninhydrin or to heat in the water bath, but each lot of ninhydrin should be checked to ensure its lack of effect. Place 5 ml. of 1% picric acid in a reaction vessel containing alundum and octyl alcohol. Boil for 1 minute, cool, evacuate, and analyze for its CO_2 content as described above. The correction, c, should not be much over 30 mm. at 0.5 ml. gas volume if the 0.5 N NaOH is prepared as directed. For a given 0.5 N NaOH solution, c is constant if the solution is guarded from atmospheric CO_2. Duplicate blank analyses are recommended with each day's analyses.

CALCULATION

The pressure, P_{CO_2}, of CO_2 from amino acid carboxyl groups is calculated as:

$$P_{CO_2} = p_1 - p_2 - c$$

mg. α-amino N per 100 ml. of plasma = $P_{CO_2} \times$ factor

The values of the factors for different temperatures are given in Table I.

NOTE 5: Small differences in temperature between blank and plasma filtrate analyses do not affect the results significantly.

NOTE 6: Where the plasma urea nitrogen concentration is within the normal limits of 5 to 20 mg. per cent, a correction of 0.1 mg. per cent may be subtracted from the α-amino nitrogen result.

TABLE I

FACTORS BY WHICH P_{CO_2} IS MULTIPLIED TO OBTAIN MILLIGRAM OF
a-AMINO NITROGEN PER 100 ML. OF PLASMA.[a]

(a = 0.5 ml.; S = 3.0 ml.; i = 1.006)[*]

Temperature	Factor	Temperature	Factor
20°C	0.0473[**]	26°C	0.0462
21°C	0.0471	27°C	0.0460
22°C	0.0469	28°C	0.0458
23°C	0.0467	29°C	0.0456
24°C	0.0465	30°C	0.0454
25°C	0.0463		

[a] Calculated from MacFadyen (6).
[*] Key: a = volume of gas at which pressure is measured. S = milliliter of solution in gas chamber from which the CO_2 is extracted. i = factor correcting for
reabsorption of CO_2 when volume is decreased from 47 to 0.5 ml.
[**] These factors apply only when the amount of plasma represented is as described in the text, and when the 0.5 N NaOH and 2 N lactic acid are dissolved
in 25% NaCl. [The factors given in Table I of reference (4) for plasma when
a = 0.5 are incorrectly calculated from Table I of reference (6)].

Procedure for Plasma of High Urea Content

Urea evolves an amount of CO_2 equivalent, in terms of a-amino
nitrogen, to 0.7% of its nitrogen. Where the urea nitrogen concentration exceeds 20 mg. per cent, the procedure outlined above requires
modification for exact results. Three alternative procedures are available: (1) the plasma may be analyzed for its urea nitrogen concentration. This value times 0.007 is subtracted from the a-amino nitrogen
result. (2) The urea may be removed with urease. Since most preparations of urease contain some a-amino nitrogen (the dialyzed preparation of Archibald and Hamilton (7) appears to be an exception), a
separate analysis of the urease must be performed. The final result is
obtained by difference, thus increasing the possibility of error. This
procedure for correcting for urea is therefore not recommended. (3)
The filtrate may be incubated with excess ninhydrin in order to bind
the urea completely to ninhydrin so that no CO_2 is evolved from the
urea. The routine procedure is followed except at the following points.
Instead of 100 mg. of ninhydrin, 200 mg. are used. The evacuated
reaction vessel is incubated for 3 hours at 60°C. or overnight at 37°C.,
and then heated for 10 or 20 minutes in the boiling water bath.

II. URINE

A complete 24-hour specimen should be collected, and the volume measured. If the analysis is not started within an hour or two after collection, the whole specimen, or an aliquot, may be preserved by saturation with thymol and storage at 4°C., or by freezing without preservative.

The urea-amino acid ratio in urine is so high that the urea must be removed with urease.

Procedure

REMOVAL OF UREA WITH UREASE

To each triplicate reaction vessels add 2 ml. of urine and 1 drop of brom thymol blue. If the reaction is alkaline (blue), add 1 N H_2SO_4 drop by drop until the solution is yellow, and then 1 N NaOH until just blue (pH a little over 6). If the urine is acid so that the indicator is yellow, add 1 N NaOH until it is just blue. When the reaction is thus adjusted, add from a calibrated glass spoon 170 ± 15 mg. of phosphate buffer of pH 6.2, 0.2 ml. of 1% urease solution and a crystal of thymol. Stopper the vessel with the opening to the side arm open, and incubate overnight at 37°–40°C. Checker N.Y.C. used the phosphate buffer solution and 2% urease. He incubated for 60 minutes at 50–52°.

REMOVAL OF CO_2, ADDITION OF NINHYDRIN, ETC.

After incubation, add 1 drop of brom cresol green and 1 drop of octyl alcohol. Add 5 N H_2SO_4 until the solution is just yellow, then 100 ± 10 mg. of citrate buffer of pH 2.5 from a calibrated glass spoon, and two or more pieces of alundum. Checker N.Y.C. added 0.5 ml. of the citrate buffer and 2 or 3 beads at this point. Boil for exactly 1 minute and proceed as described for plasma filtrate except that the time of heating in the boiling water bath is reduced from 20 to 8 minutes because of the greater concentration of ninhydrin and a slightly higher pH.

The gas volume may be read at either 0.5 or 2 ml., depending upon the amount of CO_2 evolved. The smaller gas volume is preferable if manometer readings can be obtained, but not infrequently urines evolve so much CO_2 that a 2 ml. gas volume must be used.

BLANK ANALYSIS FOR c CORRECTION

The blank, which should be run in triplicate, consists of 2 ml. of water treated exactly as the 2 ml. of urine. The c correction is p_1-p_2, with the manometer readings being made at the same gas volume as is used for the urine. Because of the presence of a-amino nitrogen in most urease preparations, the value of c in the analysis of urine is considerably higher than in plasma analysis.

CALCULATION

The pressure, P_{CO_2}, of CO_2 from amino acid carboxyl groups is calculated as:

$$P_{CO_2} = p_1 - p_2 - c$$

mg. a-amino nitrogen per 24 hour urine = P_{CO_2} × factor × $\dfrac{\text{24 hour volume in ml.}}{2}$

The values of the factors are given in Table II.

TABLE II

FACTOR BY WHICH P_{CO_2} IS MULTIPLIED TO OBTAIN MILLIGRAM OF a-AMINO NITROGEN[a]

a = 0.5[b] i = 1.006 S = 3.0	a = 2.0[b] i = 1.003 S = 3.0
Temperature	

Temperature		
20°C	0.0003938[c]	0.001571[c]
21°C	0.0003922	0.001565
22°C	0.0003906	0.001559
23°C	0.0003890	0.001552
24°C	0.0003875	0.001546
25°C	0.0003861	0.001540
26°C	0.0003846	0.001534
27°C	0.0003831	0.001528
28°C	0.0003816	0.001523
29°C	0.0003802	0.001517
30°C	0.0003787	0.001511

[a] From MacFadyen (6).
[b] Abbreviations as in Table I.
[c] Factors apply only when the 0.5 N NaOH and 2.0 N lactic acid are made up in 25% NaCl solution.

DISCUSSION

The method is capable of good precision and accuracy. Hamilton and Van Slyke (4) obtained reproducibility between duplicates of 1% or less and recovery of added amino acids of 98–101%, while corresponding values obtained by Woodruff and Man (8) were 2.4% at the most for reproducibility, with recovery of 98–102%. The writer frequently has obtained duplication comparable to that of the above authors; but, because of occasional greater discrepancies, triplicate determinations are run whenever sufficient material is available.

Normal Values

PLASMA

The largest series reported is that of Woodruff and Man (8). In thirty-seven normal adults in the postabsorptive state, the mean value was 4.23 mg. per cent, and the range 3.37–4.97 (not corrected for urea). No sex difference was noted. Other smaller series, including that of the writer, are in agreement with that of Woodruff and Man. One series on sixteen normal children (9) indicate that the lower limit may be below that of normal adults. For several hours after a high-protein meal, the values may be 1–2 mg. per cent above the fasting level (10).

Munro and Thomson (11), using a colorimetric method (1) for the determination of plasma amino nitrogen found a mean value of 6 mg. per cent in six normal fasting adults, indicating that substances in addition to free amino acids were being measured.

URINE

In *adults,* the values range from about 80–200 mg. of α-amino nitrogen per 24 hours (12, 13, 14). The data of Thompson and Abdulnabi (13) suggest that the urinary excretion is higher in females than in males. Except where excessive amounts of protein are eaten, the urinary excretion of amino acids is not affected by diet (15). In children, the mean urine α-amino nitrogen level is about 1 mg. per pound of body weight per day (9, 16).

Albanese and Irby (17), using their copper method, have found the urinary excretion of amino nitrogen in normal adults to vary between 221 and 696 mg. per 24 hours, indicating the lack of specificity of this method. The recent modification of their method (2) appears to effect some improvement.

Pathological Values

PLASMA

The α-amino nitrogen concentration of fasting human plasma falls outside the normal limits in very few diseases. In acute yellow atrophy of the liver very high levels may be reached, whereas in less severe forms of liver disease the levels are normal or only slightly elevated (9). Luetscher (18) found markedly elevated levels in his twelve patients with severe untreated diabetes mellitus. Gray and Illing (19), on the other hand, found normal plasma α-amino nitrogen levels in their fifty-three cases of diabetes mellitus. In three cases of diabetes in acidosis the writer found two to have normal values, while one was very slightly elevated. Hypoaminoacidemia has been reported (10) in children in nephrotic crisis.

URINE

There are two general causes for increased urinary excretion of amino acids: (1) elevated plasma levels, as in the diseases mentioned above; and (2) specific renal tubular defects, where aminoaciduria occurs in the presence of normal plasma levels. Examples of the latter group are Wilson's disease (20), de Toni-Fanconi syndrome (21), and galactosemia (22). The markedly increased excretion of cystine, arginine, and lysine in cystinuria (23) will probably be reflected in an increase in total α-amino nitrogen. The administration of adrenocorticotropic hormone causes an increased urinary excretion of amino acids (24).

REFERENCES

1. (a) Frame, E. G., Russell, J. A., and Wilhelmi, A. E., The colorimetric estimation of amino nitrogen in blood. *J. Biol. Chem.* 149, 255–270 (1943). (b) Russell, J. A., Note on the colorimetric determination of amino nitrogen. *J. Biol. Chem.* 156, 467–468 (1944).
2. Sobel, C., Henry, R. J., Chiamori, N., and Segalove, M., Determination of α-amino nitrogen in urine. *Proc. Soc. Exptl. Biol. Med.* 95, 808–813 (1957).
3. Van Slyke, D. D., Dillon, R. T., MacFadyen, D. A., and Hamilton, P. B., Gasometric determination of carboxyl groups in free amino acids. *J. Biol. Chem.* 141, 627–699 (1941).
4. Hamilton, P. B., and Van Slyke, D. D., The gasometric determination of

free amino acids in blood filtrates by the ninhydrin-carbon dioxide method. *J. Biol. Chem.* **150**, 231–250 (1943).

5. Van Slyke, D. D., MacFadyen, D. A., and Hamilton, P. B., The gasometric determination of amino acids in urine by the ninhydrin-carbon dioxide method. *J. Biol. Chem.* **150**, 251–258 (1943).

6. MacFadyen, D. A., Determination of amino acids in plasma by the ninhydrin-carbon dioxide reaction without removal of proteins. *J. Biol. Chem.* **145**, 387–403 (1942).

7. Archibald, R. M., and Hamilton, P. B., Removal of canavanine from preparations of Jack Bean urease. *J. Biol. Chem.* **150**, 155–158 (1943).

8. Woodruff, C. W., and Man, E. B., Concentration of α-amino acid nitrogen in plasma of normal subjects. *J. Biol. Chem.* **157**, 93–97 (1945).

9. Hsia, D. Y., and Gellis, S. S., Amino acid metabolism in infectious hepatitis. *J. Clin. Invest.* **33**, 1603–1610 (1954).

10. Farr, L. E., and MacFadyen, D. A., Hypoaminoacidemia in children with nephrotic crisis. *Am. J. Diseases Children* **59**, 782–792 (1940).

11. Munro, H. N., and Thomson, W. S. T., Influence of glucose on amino acid metabolism. *Metabolism* **2**, 354–361 (1953).

12. Cooper, A. M., Eckhardt, R. D., Faloon, W. W., and Davidson, C. S., Investigation of the aminoaciduria in Wilson's disease. *J. Clin. Invest.* **29**, 265–278 (1950).

13. Thompson, R. C., and Abdulnabi, M., A study of the urinary excretion of α-amino nitrogen and lysine by humans. *J. Biol. Chem.* **185**, 625–628 (1950).

14. Frame, E. G., and Rausch, V. L., Chromatographic studies of urinary amino acids. *Bull. Univ. Minn. Hospitals* **23**, 532–543 (1952).

15. Eckhardt, R. D., and Davidson, C. S., Urinary excretion of amino acids by a normal adult receiving diets of varied protein content. *J. Biol. Chem.* **177**, 687–695 (1949).

16. Childs, B., Urinary excretion of free α-amino nitrogen by normal infants and children. *Proc. Soc. Exptl. Biol. and Med.* **81**, 225–226 (1952).

17. Albanese, A. A., and Irby, V., Determination of urinary amino nitrogen by the copper method. *J. Biol. Chem.* **153**, 583–588 (1944).

18. Luetscher, J. A., The metabolism of amino acids in diabetes mellitus. *J. Clin. Invest.* **21**, 275–279 (1942).

19. Gray, C. H., and Illing, E., Plasma and urinary amino acids in diabetes. *J. Endocrinol.* **8**, 44–49 (1952).

20. Uzman, L., and Denny-Brown, D., Aminoaciduria in hepato-lenticular degeneration (Wilson's disease). *Am. J. Med. Sci.* **215**, 599–611 (1948).

21. McCune, D. J., Mason, H. H., and Clarke, H. T., Intractable hypophosphatemic rickets with renal glycosuria and acidosis (Fanconi syndrome). *Am. J. Diseases Children* **65**, 81–146 (1943).

22. Hsia, D. Y., Hsia, H., Green, S., Kay, M., and Gellis, S. S., Aminoaciduria in galactosemia. *Am. J. Diseases Children* **88**, 458–465 (1954).

23. Stein, W. H., Excretion of amino acids in cystinuria. *Proc. Soc. Exptl. Biol. Med.* **78**, 705–708 (1951).

24. Ronzoni, E., Roberts, E., Frankel, S., and Ramasarma, G. B., Influence of administration of ACTH on urinary amino acids. *Proc. Soc. Exptl. Biol. Med.* **82**, 496–503 (1953).

TURBIDIMETRIC MEASUREMENT OF AMYLASE: STANDARDIZATION AND CONTROL WITH STABLE SERUM*

Submitted by: Arnold G. Ware, Clifford B. Walberg, and Rex E. Sterling, Chemistry Division of the Main Laboratory, Los Angeles County Hospital, Los Angeles, California, and Department of Biochemistry and Nutrition, University of Southern California, School of Medicine, Los Angeles, California

Checked by: David Seligson, Yale University, New Haven, Connecticut

Introduction

Waldron's turbidimetric method (1) for measuring amylase in pancreatic fluid has been applied to serum and/or urine by a number of investigators (2, 3, 4). The amount of substrate hydrolysis is estimated from turbidimetric readings made at constant time intervals. Enzyme activity is calculated from the data obtained by application of a formula based on kinetic principles.

This approach to measuring amylase has been found to be somewhat impractical in our laboratory. Calculation of enzyme activity is cumbersome and, in addition, the turbidity imparted by the serum specimen complicates the measurement of starch concentration. We, therefore, were attracted to a slightly modified procedure which obviates these difficulties by relating the enzyme activity directly to the decrease in turbidity produced during a constant time interval. This approach is satisfactory and it minimizes the difficulties of specimen turbidity.

In applying this technique, it occurred to us that serum with known amylase activity might be used as a daily check on enzyme-subtrate relationships. Once the turbidimetric response to known enzyme concentrations is determined with a particular substrate and photometer, it is necessary only to run a daily check at an appropriate level for validation. For this purpose we originally used frozen serum with known enzyme activity. Recently a fortified, stable, lyophilized serum

---

* Based on the method of Waldron (1) and Peralta and Reinhold (3).

15

has become available which has greatly extended the convenience of the method.

The validity of such a method is dependent upon the accuracy of turbidimetric measurements and upon the availability of a starch suspension with suitable particle size. Waldron (1) and Peralta and Reinhold (3) pointed out the usefulness of a standardized starch suspension available commercially under the trade name of "Zippy." For turbidimetric measurements both investigators used the Klett-Summerson photocolorimeter which has excellent electrical stability and permits acceptable reproducibility in photometric readings.

Reagents

1. *Zippy starch.*[1] This is a 10% suspension in 4% NaCl preserved with pine oil.

2. *Starch diluent,* pH 7.0–7.2. NaCl, 2.5 g.; $Na_2HPO_4 \cdot 12H_2O$, 17.3 g.; and KH_2PO_4, 1.6 g.; make up to 1 liter with distilled water.

3. *Working starch suspension.* Thoroughly mix the Zippy starch by inverting the bottle several times. To 1 volume of starch, add sufficient diluent (approximately 9 volumes) to give a reading of 350 scale divisions (±25) on the Klett-Summerson photocolorimeter when read against a water reference. If the reading falls outside the range indicated, adjustments in the dilution should be made. This reading on

---

[1] Zippy, Inc., Los Angeles 63, Calif. Generally available at local markets.

A. A recent change in the formulation of commercial Zippy Starch has seriously limited its usefulness in this test. The authors recommend as a substitute, "Harleco's Starch Suspension, Zippy." Hyland Laboratories, Los Angeles, has introduced a similar formulation which includes buffer and requires only dilution with distilled water. The use of this preparation is also recommended.

B. Using either of the starch preparations recommended in (A) it has been found that the standard curve assumes approximate linearity when the logarithm of the amylase activity is plotted against the decrease in absorbance. In routine use therefore, a curve may be constructed from only two dilutions of the reference serum. We use amylase levels of approximately 100 and 400 units plotted on the logarithmic scale of semilogarithmic graph paper.

C. In order to increase the sensitivity of the method, a 10 minute time interval may be used. Under these circumstances the best working range for the standard curve is between 30 and 300 amylase units.

D. If instruments are used with varying cuvette sizes, the starch dilution with buffer should be altered so that the working suspension gives absorbance readings between 0.60 and 0.70 when read against a water reference.

the Klett instrument corresponds to an absorbance of 0.70 and a transmittance of 21%. The suspension may be kept at 37°C. during the working day but should be mixed well to suspend the starch particles before removing samples. In practice, we prepare the working suspension daily.

NOTE: The working suspension may be stored in the refrigerator under which conditions it remains usable until the turbidity decreases below the range indicated. It is necessary when pipetting the working starch suspension to guard against contamination with saliva.

Procedure

1. Place 1.0 ml. of specimen to be tested in a cuvette and warm to 37°C. in a water bath or heating block.

2. Add rapidly 5.0 ml. of working starch suspension which has been prewarmed to 37°C. (zero time). Invert twice immediately, place in photometer (red filter #66, Klett) and read at exactly 0.5 minutes against a water reference. Return cuvette to the 37°C. water bath.

NOTE: It is important to use gentle inversion for mixing of the starch and sample. Vigorous shaking will incorporate air bubbles into the mixture which makes it difficult to obtain accurate photometric readings.

NOTE: The temperature of the reaction may vary from 37°C. as long as the same temperature is used for preparing the standard curve and for running unknown samples.

3. At 5 minutes, remove the cuvette from the water bath, invert twice, and read again at exactly 5.5 minutes.

NOTE: When making the photometric readings at 0.5 and 5.5 minutes, we have found it convenient to adjust the potentiometer to the approximate null point a few seconds before the reading time so that an accurate reading may be obtained at the exact time.

4. Subtract the last reading from the first to obtain the decrease in absorbance which occurred during the 5-minute period.

5. From the decrease in absorbance, obtain the amylase activity in units by referring to a standard curve or chart. If the decrease in absorbance is more than 50 Klett scale divisions, the activity exceeds 500 units and the original specimen should be diluted with water or saline and the entire procedure repeated. The activity of the diluted specimen should fall between 50 and 500 units and must be multiplied

by the dilution factor to obtain the amylase activity of the undiluted specimen.

Standardization

A serum is selected which assays 500 units or more by the Somogyi saccharogenic method, as modified by Henry and Chiamori (5). This serum is diluted with water or saline in such a way as to yield amylase levels of approximately 50, 100, 200, 300, 400, and 500 units. Each sample is then tested in triplicate in steps 1, 2, 3, and 4 of the procedure. The average decrease in Klett scale divisions for each amylase level is then plotted against the corresponding units on arithmetic graph paper. A typical standardization curve is presented in Fig. 1.

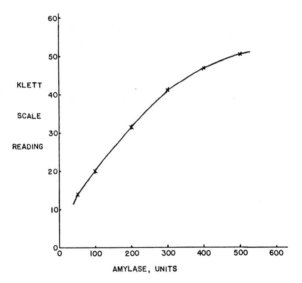

Fig. 1. Typical standardization curve relating amylase activity to decrease in turbidity.

Amylase values above 500 units are not included since it has been found that decreases in turbidity fall off rapidly above this level.

NOTE: The sodium chloride concentration of the substrate is sufficiently high so that distilled water may be used to dilute specimens. Physiological saline may also be used for dilution without altering the results.

Control

A daily check on the standard curve is carried out with a serum sample of known amylase content (200 to 400 unit range). If the results of this assay fall outside the acceptable limits (see Precision of the Method), it may be necessary to construct another standard curve. As a general rule, the curve does not vary significantly beyond these limits unless there has been a change in the substrate, instrument, or technique. The most frequently encountered changes result from shifting of the position of the light source when changing photometer bulbs.

Stability of Amylase in Serum

NATIVE SERUM

Over the past few years we have checked the amylase activity of a number of frozen sera kept at −15°C. Each serum was divided among Pyrex tubes, stoppered, frozen, and a single tube thawed when needed for assay. There was no detectable change in amylase activity in any of the sera during the period tested (5 months).

FORTIFIED, LYOPHILIZED SERUM

Crystalline alpha amylase of porcine origin[2] is added to pooled, normal, human serum to obtain an approximate level of 1000 Somogyi units per 100 ml. The serum is then accurately measured into glass bottles, frozen, lyophilized, stoppered, and sealed by standard techniques.[3] The bottles are opened, reconstituted with distilled water, and assayed. We have found no detectable change in amylase activity in the lyophilized product stored at room temperatures for 2 months. When kept at refrigerated temperatures, no detectable change occurred over a period of 3 years.

The reconstituted serum shows no significant change in amylase activity for periods up to 5 days when kept at 4°C. If kept in the frozen state, the reconstituted serum is stable for at least 3 weeks. Repeated freezing and thawing should be avoided, however.

[2] Worthington Biochemical Corp. Purified according to the method of Caldwell et al. (6).

[3] Available as Abnormal Clinical Chemistry Control Serum, Hyland Laboratories, Los Angeles 39, California.

Accuracy of the Method

The accuracy of the method is directly related to that of the reference method used to standardize the serum. The use of fortified serum for standardization does not appear to affect the accuracy of the method when applied to serum or urine.

Thirty-six samples of fortified sera were assayed by both the turbidimetric and saccharogenic methods.[4] The turbidimetric method was standardized as outlined above, with pooled serum from patients with acute pancreatitis. The amylase level of this pooled serum was determined by the saccharogenic method. Amylase levels of the fortified sera varied between 580 and 1600 units. All of the results with the two methods agreed within 20%. Statistical analysis indicates that a significant difference between the results of the two methods did not occur.

Diastase levels were determined on ten urines by the turbidimetric method and by an amyloclastic method.[5] The turbidimetric method, in this instance, was standardized with pooled serum assayed by the same amyloclastic method using photometric measurement of the starch-iodine end point (7). The urines varied in amylase activity between 100 and 9600 units per 100 ml. All of the results of the two methods agreed within 20%. Statistical analysis indicates that a significant difference between the results of the two methods did not occur.

Precision of the Method

The precision of the method is dependent upon the photometer used. With the Klett-Summerson photocolorimeter we have found the reproducibility (2 standard deviations) to be approximately ±2 Klett scale divisions. This represents an amylase variation of approximately ±15%, with the percentage increasing somewhat at the lower part of the scale. Because of lack of precision at low levels, we do not report values below 50 amylase units.

With direct reading instruments such as the Coleman Junior Spectrophotometer, the precision decreases because the photometric readings are comparatively less reproducible. Therefore, we recommend the use of a null reading instrument such as the Klett-Summerson photocolorimeter.

[4] Assays by the saccharogenic method were performed at BioScience Laboratories, Los Angeles, California.

[5] The amyloclastic method was selected because it was in use in our laboratory at the time this experiment was performed.

Turbid Sera

When the turbidity of a specimen is such that it elevates the original reading (Procedure, step 2, 0.5 minutes) to a value higher than that of the working starch suspension, the accuracy of the method may be affected. With highly lactescent specimens, the results are markedly inaccurate. However, it is possible to obtain accurate results on these specimens after extraction with diethyl ether. The specimen is shaken vigorously with 2 volumes of ether and then centrifuged. The clear subnatant may then be used for the amylase determination. We have found no significant change in amylase activity with many non-lactescent specimens treated in this manner. If the specimen contains a large amount of fat, the amylase activity per unit volume may actually increase after the fat is removed.

Normal Values

Using the saccharogenic methods of Henry and Chiamori (5) for standardization of the procedure described, normal levels are as follows:

Serum: 50 to 150 units per 100 ml.

Urine: less than 500 units per 100 ml. or less than 200 units per hour.

These levels are estimated to include 95% of the normal population.

REFERENCES

1. Waldron, J. M., Photometric determination of amylase activity in pancreatic juice. *J. Lab. Clin. Med.* **38**, 148–152 (1951).
2. Jacobson, L., and Hansen, H., Turbidimetric determination of diastase in urine with dextrin as substrate. *Scand. J. Clin. Lab. Invest.* **4**, 134–141 (1952).
3. Peralta, O., and Reinhold, J. G., Rapid estimation of amylase activity of serum by turbidimetry. *Clin. Chem.* **1**, 157–164 (1955).
4. Guth, P. H., Evaluation of phototurbidimetric technics for the determination of serum amylase, lipase and esterase. *Am. J. Gastroenterol.* **33**, 319–334 (1960).
5. Henry, R. J., and Chiamori, N., Study of the saccharogenic method for the determination of serum and urine amylase. *Clin. Chem.* **6**, 434–452 (1960).
6. Caldwell, M. L., Adams, M., Kung, J., and Toralballa, G. C., Crystalline pancreatic amylase. II. Improved method for its preparation from hog pancreas glands and additional studies of its properties. *J. Am. Chem. Soc.* **74**, 4033 (1952).
7. Van Loon, E. J., Likins, M. R., and Segar, A. J., Photometric method for blood amylase by use of starch-iodine color. *Am. J. Clin. Pathol.* **22**, 1134–1136 (1952).

IDENTIFICATION OF URINARY TRACT CALCULI BY INFRARED SPECTROSCOPY*

Submitted by: BERNARD KLEIN, Veterans Administration Hospital, Bronx, New York

Checked by: HARRIET T. SELIGSON, Yale University School of Medicine, New Haven, Connecticut

Introduction

Precise knowledge of the composition of urinary tract calculi is required to determine the possible cause of stone formation and also to plan a program of medical management to prevent recurrence. The analyst must often work with very small stones or fragments the size of which precludes tedious quantitative analysis. Qualitative analytical schemes have been devised but are sometimes inconclusive.

The increasing availability of infrared spectrophotometers coupled with their capacity for analyzing nondestructively, milligram amounts of organic and inorganic substances has led to the successful application of infrared spectrophotometry to the identification of renal tract calculi (1). The potassium bromide pellet sample preparation has even further simplified the procedure (2, 3).

Reagents

1. *Potassium bromide.* Analytical grade material is ground to about 40 mesh and heated in an oven at 120°C. for 3 hours. The dried salt is cooled in a desiccator, bottled in 5 gm. portions and stored in a desiccator.

NOTE: Infrared quality potassium bromide may be obtained from Harshaw Chemical Co., Cleveland 6, Ohio.

Equipment

1. *Infrared recording spectrophotometer.* Equipped with sodium

---

* Based on the method of Weissman et al. (1).

chloride optics, having a scanning span from 2.0 to 15.0 μ. The instrument is calibrated with a polystyrene film.

NOTE: In the submitter's laboratory, infrared absorption spectra were obtained with a Perkin-Elmer Model 21 Recording Spectrophotometer.

2. *Die, vacuum type.* For preparation of potassium bromide pellets.

NOTE: Proper care of the die is essential since corrosion is intensified by the use of the salt. Optical polish of the die faces will help reduce imperfections in the plate which may produce scattering losses. It is also helpful to rub lightly the surface of the split die with a good grade of paraffin wax, from time to time.

3. *Press, laboratory type.* The press should be capable of delivering at least 20,000 lbs./sq. in.

4. *Pump, vacuum,* oil. Capable of producing 0.1 mm.

5. *Holder,* pellet. For infrared spectrophotometer.

Procedure

The calculus, whenever possible, is sectioned and if layers are visible, they are separated and examined individually. Otherwise, the calculus is crushed in an agate or mullite mortar and ground as fine as possible.

A potassium bromide pellet is prepared:

(1) 1–5 mg. finely ground sample is mixed intimately with 250 mg. dried KBr and the mixture is poured into a cleaned, dry die.

NOTE: A calibrated glass measuring spoon prepared as described by Van Slyke and Folch (4) is convenient for measuring KBr powder.

NOTE: A very satisfactory method for grinding and mixing the sample and KBr is to place both in a polyethylene capsule with a stainless steel ball pestle and to shake the mixture in a mechanical vibrator grinder used by dentists to prepare amalgams. (Wig-L-Bug amalgamator, Crescent Dental Mfg. Co., Chicago, Ill.)

(2) The powdered surface is leveled and the die ram is inserted. The entire die is assembled and placed in the press.

(3) The die is evacuated for about 5 minutes before pressure is applied. Pressure is then gradually increased to about 20,000–22,000 lbs./sq. in. and maintained for the length of time required to produce clear pellets. This is usually 3–5 minutes.

NOTE: A good quality pellet should be completely transparent, with small specks of the sample interspersed in the matrix. Opaque areas indicate that moisture has not been completely removed. In that case, the pellet should be broken up and the pelleting process repeated.

(4) Pressure and vacuum are released, in that order and the pellet is carefully extracted from the die, placed in the pellet holder and the spectrum is obtained.

NOTE: The spectrum is recorded in the manner described for the instrument in use. In the submitter's laboratory, the programming arrangement was: Program 925, autosuppression, 0, gain 4–5, response 1, speed 4. The gears on the drum drive were arranged to record the spectrum on an 8½ × 11-inch piece of millimeter ruled graph paper. This is very convenient for filing and facsimile reproduction. The resulting spectrum is studied for characteristic absorption peaks or bands. These are compared to reference spectra and classified as to composition.

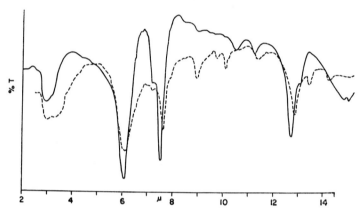

FIG. 1. Reference spectra of calcium oxalate. Characteristic absorptions group: hydroxyl, carboxyl. Wavelength (μ): 2.85–3.0; 6.10, 7.5, 12.7 (sharp).

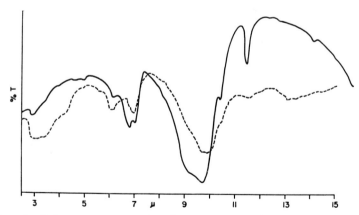

FIG. 2. Reference spectra of carbonato-apatite. Characteristic absorptions group; carbonate, phosphate. Wavelength (μ): 6.85, 7.05, 11.4; 9.1–9.6 (broad).

B. KLEIN

Reference Spectra

Reference spectra are prepared for the commonly occurring calculi
and mixtures. They are calcium oxalate, calcium hydroxy/carbonato-
phosphate (apatite), magnesium ammonium phosphate ("triple
phosphate"), uric acid, and cystine, and are shown in Figs. 1–9. The
dotted lines show actual spectra.

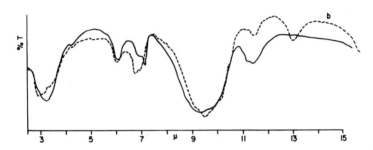

FIG. 3. Reference spectra of magnesium ammonium phosphate. Characteristic
absorptions group: ammonium ion, phosphate. Wavelength (μ): 3.25; 9.0–10.0
(broad).

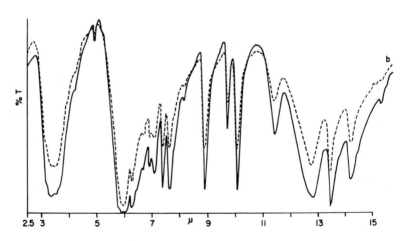

FIG. 4. Reference spectra of uric acid. Characteristic absorptions group:
amino N-H, carbonyl. Wavelength (μ): 3.5; 5.9–6.2. Note the typical complex
organic spectrum.

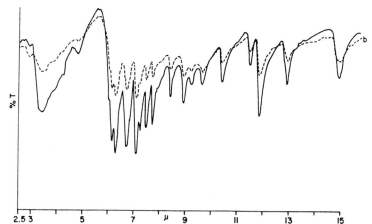

Fig. 5. Reference spectra of cystine. Characteristic absorptions group: ammonium ion (zwitterion form), ammonium ion (deformation), carboxyl ion. Wavelength (μ): 3.3; 6.35; 6.7; 6.15. Note the complex organic spectrum.

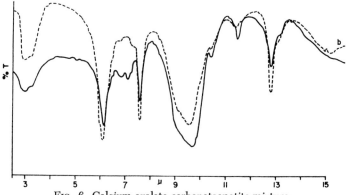

Fig. 6. Calcium oxalate-carbonatoapatite mixture.

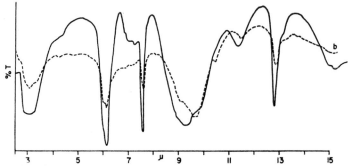

Fig. 7. Magnesium ammonium phosphate-calcium oxalate mixture.

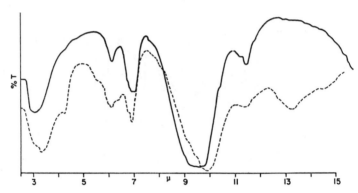

Fig. 8. Magnesium ammonium phosphate-carbonatoapatite mixture.

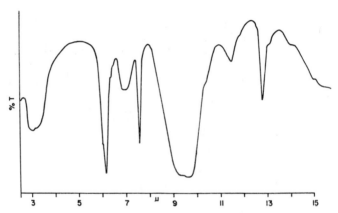

Fig. 9. Magnesium ammonium phosphate-calcium oxalate-carbonatoapatite mixture.

Discussion

The potassium bromide pellet is a more convenient sampling method than the Nujol or fluorolube mulling method for solid state analysis. Further, no interfering absorptions of the mulling medium are encountered. Interaction of the KBr with the sample has been reported, but this has not been observed in the examination of inorganic calculi. The submitter has encountered additional and spurious absorption bands when calcium carbonate is incorporated into a KBr pellet. This solid solution form of dispersion is subject, of course, to the complications of polymorphism and orientation effects. A brief but valuable discussion of infrared absorption analysis of powdered samples is given by Butz (5). Another advantage of the pellet technique is that the pellet can be stored in a desiccator for future reference or reexamination.

A pure, single component, inorganic calculus is an infrequent occurrence. Preparation of reference spectra taken on mixtures of known composition can give the analyst a reliable semiquantitative estimate of the constituents in the calculus.

Applications of infrared spectroscopy to the identification of other pathological concretions has been reported (6, 7).

REFERENCES

1. Weissman, M., Klein, B., and Berkowitz, J., Clinical applications of infrared spectroscopy. Analysis of renal tract calculi. *Anal. Chem.* **31**, 1334-1338 (1959).
2. Stimson, M., and O'Donnell, M. J., The infrared and ultraviolet absorption spectra of cytosine and isocytosine in the solid state. *J. Am. Chem. Soc.* **74**, 1805-1808 (1952).
3. Scheidt, U., and Reinwein, H., The infrared spectroscopy of amino acids. *Z. Naturforsch.* **76**, 270-277 (1952).
4. Van Slyke, D. D., and Folch, J., Manometric carbon determination. *J. Biol. Chem.* **136**, 509-541 (1940).
5. Butz, W. H., Beckman Instruments Division, Application Data Sheet, IR 8063.
6. Klein, B., Weissman, M., and Berkowitz, J., Clinical applications of infrared spectroscopy. II. Identification of pathological concretions and other substances. *Clin. Chem.* **5**, 453-465 (1960).
7. Beckman Application Data Sheet, IR-8074-M. Beckman Instruments, Inc. Fullerton, California.

ADDITIONAL REFERENCES

Bellamy, L. J., "The Infrared Spectra of Complex Molecules," 2nd Ed. Wiley, New York, 1958.

Jones, R. M. and Sandorfy, C., The application of infrared and raman spectrometry to the elucidation of molecular structure, In "Chemical Applications of Spectroscopy," (W. West, ed.) 247-563. Interscience, New York, 1956.

THE SPECTROPHOTOMETRIC DETERMINATION OF CARBON MONOXIDE IN BLOOD*

Submitted by: JOSEPH S. AMENTA, Yale University, New Haven, Connecticut
and Walter Reed Army Medical Center, Washington, D.C.

Checked by: R. P. McDONALD, Harper Hospital, Detroit, Michigan
A. HAINLINE, JR., Cleveland Clinic, Cleveland, Ohio
D. McKAY, Grace-New Haven Community Hospital, New Haven, Connecticut

Introduction

The spectrophotometric determination of blood carbon monoxide, first developed by Hufner in 1900 (5), is a simple, rapid, and reliable technique well suited to the needs of the clinical laboratory. The basis of this technique lies in the observation that the spectral absorbance curves of oxyhemoglobin and carboxyhemoglobin are different (Fig. 1). Hufner measured the difference in absorbance between the two

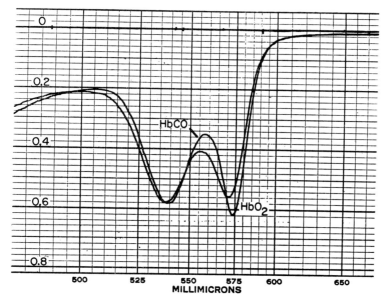

FIG. 1. Spectral-absorbance curves of carboxyhemoglobin and oxyhemoglobin. The ordinate is absorbance; the abscissa is wavelength.

* Based on methods of Hufner (5), Heilmeyer (5), and Drabkin (1, 3).

31

forms of hemoglobin at wavelengths 541 and 560 mμ and used the absorbances differences to measure oxy- and carboxyhemoglobin. The modifications that have since been developed are basically of two types: (a) utilization of two wavelengths other than 541 and 560 mμ originally suggested by Hufner and (b) the algebraic summation of extinction coefficients at more than two wavelengths as recommended by Drabkin (1).

Principle

At any wavelength, x, the extinction coefficient of oxyhemoglobin is defined:

$$E_{x1} = \frac{A_{x1}}{HbO_2} \qquad (1)$$

where E_{x1} is the extinction coefficient of oxyhemoglobin at wavelength x, A_{x1} is absorbance and HbO_2 is the concentration of oxyhemoglobin in grams per 100 ml. of blood. Length of the light path is taken as 1 cm. Similarly, for carboxyhemoglobin at wavelength x:

$$E_{x2} = \frac{A_{x2}}{HbCO} \qquad (2)$$

Since absorbances of two pigments are additive, the total absorbance of a mixture of oxyhemoglobin and carboxyhemoglobin at wavelength x is:

$$A_x = A_{x1} + A_{x2} \qquad (3)$$

where A_x is the total absorbance of the mixture. From Equations 1, 2, and 3, it follows that:

$$A_x = E_{x1}(HbO_2) + E_{x2}(HbCO) \qquad (4)$$

Similarly, it can be derived that at any other wavelength, y,

$$A_y = E_{y1}(HbO_2) + E_{y2}(HbCO) \qquad (5)$$

where A_y is the total absorbance at y, E_{y1} is the extinction coefficient of oxyhemoglobin at y, and E_{y2} is the extinction coefficient of carboxy-hemoglobin at y. These extinction coefficients are determined by experiment from standard solutions of oxyhemoglobin and carboxy-hemoglobin. By the determination of the total absorbance at each of the two wavelengths, A_x and A_y the above simultaneous equations, 4 and 5, containing two unknown quantities, HbO_2 and HbCO, can be solved by graphic or algebraic methods. For the algebraic solutions, $R_{x/y}$ is defined:

$$R_{x/y} = \frac{A_x}{A_y} \qquad (6)$$

From 4, 5, and 6

$$R_{x/y} = \frac{E_{x1}(HbO_2) + E_{x2}(HbCO)}{E_{y1}(HbO_2) + E_{y2}(HbCO)} \qquad (7)$$

by definition

$$(Hb) = (HbO_2) + (HbCO) \qquad (8)$$

where (Hb) is the total hemoglobin concentration

$$\therefore R_{x/y} = \frac{E_{x1}(Hb) - E_{x1}(HbCO) + E_{x2}(HbCO)}{E_{y1}(Hb) - E_{y1}(HbCO) + E_{y2}(HbCO)} \qquad (9)$$

$$R_{x/y} = \frac{(E_{x2} - E_{x1})(HbCO) + E_{x1}(Hb)}{(E_{y2} - E_{y1})(HbCO) + E_{y1}(Hb)} \qquad (10)$$

Since only a relative measurement of carboxyhemoglobin is desired, that is, per cent of total hemoglobin, (Hb) can be given the arbitrary value of 1. Gordy and Drabkin (3) have suggested the further simplification of selecting wavelength y, so that $E_{y2} = E_{y1}$, i.e., wavelength y is an isosbestic point. This results in:

$$R_{x/y} = \frac{(E_{x2} - E_{x1})}{E_{y1}} (HbCO) + \frac{E_{x1}}{E_{y1}} \qquad (11)$$

which is the form of the equation of a straight line; $y = mx + b$. It can be similarly derived as shown above that for any other wavelength, z,

$$R_{z/y} = \frac{A_z}{A_y} \qquad (12)$$

which is also a linear function of carboxyhemoglobin. These linear expressions 11 and 12 can be subtracted to give still another linear function of carboxyhemoglobin:

$$R = \frac{A_x - A_z}{A_y}; \quad R = R_{x/y} - R_{z/y} \tag{13}$$

By this means, more than two wavelengths can be used in each determination of carboxyhemoglobin, as suggested by Drabkin (1); these additional readings should increase the accuracy of the method.

In the method described in this chapter blood is added to the dilute ammonia solution and absorbances determined at 575, 560, and 498. The ratio obtained:

$$R = \frac{A_{575} - A_{560}}{A_{498}}$$

is compared with ratios obtained with standard solutions of oxyhemoglobin and carboxyhemoglobin and the percentage of carboxyhemoglobin calculated.

NOTE: Since 498 mμ is an isosbestic point, small errors in the wavelength setting will not greatly affect the absorbance reading. Wavelength 575 mμ and 560 mμ are approximate maximum and minimum points for the spectral absorbance curve of oxyhemoglobin which can be accurately set on the scales of most spectrophotometers. The analyst can select wavelengths of his own choice. As shown in the above derivation, a linear relationship between (HbCO) and R is obtained only when the absorbance in the denominator is determined at an isosbestic point.

Reagents

1. NH₃, 0.007 N. Dilute 0.48 ml. of concentrated 28% NH₃ to 1000 ml. with water.

NOTE: The concentration of this alkaline solution is not critical. Harboe (4) has shown that the rate of formation of alkaline hematin increases with increasing alkalinity. In order to avoid alkaline hematin formation, a dilute NH₃ solution is used.

2. HbO₂ and HbCO standards. Blood collected in the usual anticoagulants (EDTA, oxalate) can be used. Blood for standards should be obtained from nonsmokers, since smokers may have up to 7% of hemoglobin as carboxyhemoglobin. A portion of the sample is equilibrated with O₂ preferably, or with air. This can be done by placing a 5 ml. aliquot of blood into a 125 ml. separatory funnel containing

O_2 or air and rotating gently for 15 minutes. Another portion is similarly equilibrated with CO, obtained from commercial coal gas or by addition of sulfuric acid to warm formic acid (6).

NOTE: Checker R. P. McD. noted that the above directions such as using 5 ml. of blood should be followed exactly. He also preferred using pooled blood before making the HbO_2 and HbCO standards.

Apparatus

A spectrophotometer with a band pass of 5 mμ or less is required to achieve the necessary sharpness in the spectral absorbance peaks. We have performed this method with the Perkin-Elmer 4000A recording spectrophotometer and the Beckman Model B spectrophotometer. The ratios of absorbances obtained for the standard solutions differed slightly because of the better resolution of the former instrument. Each spectrophotometer, therefore, requires standardization. Dilution can be varied to suit the light path of the cuvettes available. For the dilutions used in this procedure, a 1.00 cm. cuvette was found adequate.

NOTE: Readings taken on a recording spectrophotometer simplify the method even further.

Procedure

Add 20.0 μl. of blood to a 10-ml. test tube containing 4.0 ml. of 0.007 N NH_3. Invert three times. Transfer a portion of this solution to a 1.00-cm. cuvette and read absorbance immediately at wavelengths 575, 560, and 498 mμ, using 0.007 N NH_3 as the blank or reference.

NOTE: Blood is obtained from a finger-tip or a tube of blood obtained by venipuncture. This tube should be stoppered if the analysis is delayed.

Standard Curve

The HbO_2 and HbCO standards are treated exactly as the unknown samples above. Define the spectral absorbance curve on the instrument used, covering the areas around 575, 560, and 498 to select the isosbestic point near 498 and the points of maximum difference at 575 and 560 which can be accurately set on the spectrophotometer. Slight deviation from 498 mμ may be necessary to locate the exact isosbestic point on a particular instrument.

If desired, the linear relationship between absorbance ratios and carboxyhemoglobin concentration can be checked by mixing appropriate amounts of the two standards described above and measuring these mixtures in the spectrophotometer. For example, a mixture containing 1.00 ml. of HbCO standard and 4.00 ml. of HbO_2 standard should give a ratio which calculates to 20%. Similarly, mixtures containing 40, 60, and 80% of HbCO standard can be prepared and tested.

Calculation

After measurement of absorbances (A_{575}, A_{560}, A_{498}), the carboxyhemoglobin is calculated as follows:

1. Ratio for HbO_2 standard:

$$R_{O_2} = \frac{A_{575} - A_{560}}{A_{498}}$$

2. Ratio for HbCO standard:

$$R_{CO} = \frac{A_{575} - A_{560}}{A_{498}}$$

3. Ratio for unknown:

$$R_x = \frac{A_{575} - A_{560}}{A_{498}}$$

4. Percentage of HbCO:

$$= \frac{R_{O_2} - R_x}{R_{O_2} - R_{CO}} \times 100$$

Discussion

R_{O_2} obtained with the Beckman Model B was 1.097 ± 0.004 for four determinations and R_{CO} was 0.057 ± 0.009 for four determinations. Standard samples, containing varying proportions of HbCO and HbO_2 showed a maximum random variation of 3% and conformed to the linear relationship as defined. For optimum accuracy, it is recommended that 40 μl. instead of 20 μl. be used with severely anemic patients. Since the extinction coefficient of HbCO and HbO_2 are the same at 498 mμ, the absorbance at this wavelength can also be used to determine total hemoglobin concentration. On our instrument, using the above procedure 1 + 200 dilution:

$$G\% \text{ hemoglobin} = A_{498} \times 61.2$$

Hemoglobin changes to alkaline hematin in alkaline solutions. This effect is minimized by using a dilute NH_3 solution. Although immediate reading of the diluted Hb is advised, the spectral absorbance curve has been found to be stable up to 30 minutes without significant change. Other abnormal forms of hemoglobin (methemoglobin and sulfhemoglobin) will interfere with this method. These, however, are rarely encountered in the usual experience in sufficient concentration to interfere with the HbCO determination. If a form of methemoglobin is suspected, an absorbance peak will be found in the region of 630 mμ and its presence in this circumstance, the method cannot be applied.

This method is rapid and simple. With the exception of the spectrophotometer, minimal equipment and reagents are needed. These advantages make this procedure well suited for a test that is not performed daily in the clinical laboratory and which is more often than not an emergency procedure. When a calibrated photometer is available, the method can be performed in less than 5 minutes. The method is flexible so that other wavelengths than those used above may be used if the analyst so desires. More readings at different wavelengths can be taken if greater accuracy is desired. Changes in the concentration of the NH_3 solution are not critical. Blood samples which are tightly stoppered and refrigerated are stable for 24–48 hours. The standard curve is a straight line; the entire curve can rapidly be checked by determining the end points of the line with any available blood specimen by aerating one aliquot and by adding CO to another aliquot.

An alternative method reduces the HbO_2 and the HbCO by the addition of $Na_2S_2O_3$ (7). Although this method gives greater absorbance differences between the two types of Hb, the spectral absorbance curves were less stable, thereby producing variation in the standard curves.

In patients in this hospital in whom death was attributed to CO by history and post-mortem examination, HbCO values from 50 to 87% were found. Clinically apparent symptoms are noted at much lower concentrations, in chronic CO poisoning. In some instances, these chronically ill patients may cause difficult diagnostic problems (2) until the poisoning is detected.

REFERENCES

1. Drabkin, D. L., Photometry and spectrophotometry. In "Medical Physics", (Otto Glasser, ed.). pp. 967–1008. Year Book Publishers, Chicago, 1944.
2. Gilbert, G. J. and Glaser, G. H., Neurologic manifestations of chronic carbon monoxide poisoning. New Engl. J. Med. 261, 1217–1220 (1959).
3. Gordy, E. and Drabkin, D. L., Spectrophotometric studies, XVI. Determination of the oxygen saturation of blood by a simplified technique, applicable to standard equipment. J. Biol. Chem. 227, 285–299 (1957).
4. Harboe, M., Studies on the conversion of oxyhemoglobin to alkaline hematin in dilute alkaline solutions. Scand. J. Clin. Lab. Invest. 11, 138–142 (1959).
5. Hufner, G., quoted in L. Heilmeyer, "Spectrophotometry in Medicine," Adam Hilger, London, 1943.
6. Natelson, S. and Menning, C. M., Improved methods of analysis for oxygen, carbon monoxide, and iron on finger-tip blood. Clin. Chem. 1, 165–177 (1955).
7. Klendshoj, N. C., Feldstein, M. and Sprague, A. L., The spectrophotometric determination of carbon monoxide. J. Biol. Chem. 183, 297–303 (1950).

CERULOPLASMIN ASSAY IN SERUM: STANDARDIZATION OF CERULOPLASMIN ACTIVITY IN TERMS OF INTERNATIONAL ENZYME UNITS*

Submitted by: EUGENE W. RICE, W. H. Singer Memorial Research Laboratory of the Allegheny General Hospital, Pittsburgh, Pennsylvania

Checked by: EDWARD WAGMAN, Veterans Administration Hospital, West Haven, Connecticut

YASUO TAKENAKA, Children's Hospital, Honolulu, Hawaii

Introduction

In normal human blood plasma about 90–95% of the copper exists in the form of a cuproprotein termed "ceruloplasmin" (1). The remainder of the plasma copper is loosely bound to other proteins, chiefly albumin. Ceruloplasmin is a blue alpha$_2$-globulin of plasma with a molecular weight of 151,000 and containing about 0.34% copper, which corresponds to eight atoms of copper per molecule. Ceruloplasmin contains hexosamine, hexose, and neuraminic acid. Although the physiological function of plasma ceruloplasmin is unknown, its *in vitro* enzymatic activity as an oxidase toward certain polyphenols and polyamines, particularly *p*-phenylenediamine (PPD) is well known. Ceruloplasmin, like hemoglobin, albumin, haptoglobin, and transferrin, is heterogeneous, but the different human fractions appear to be indistinguishable with respect to certain characteristic properties such as color, copper content, and oxidase activity (1, 2).

Many quantitative methods for ceruloplasmin in plasma (or serum) have been proposed which are based on four principles: by measurement of the absorption of light of 605 mμ by ceruloplasmin before and after destruction of its blue color; by measurement of the total copper after removal of free or loosely bound copper; by measurement of the rate of oxidation of *p*-phenylenediamine, or similar compounds, in a Warburg apparatus or a spectrophotometer; and by immunochemical precipitation of ceruloplasmin and quantitative estimation of the specific precipitate. Various simpler qualitative and

---

* Based on the method of Houchin (4) as modified and standardized by Rice (3, 6).

semiquantitative modifications of these methods have been devised, often by combining the method with electrophoretic analysis. Pertinent articles on ceruloplasmin methodology are listed in references (1) and (3).

Most of the published methods for the determination of serum oxidase (ceruloplasmin) activity have made no attempt to demonstrate a satisfactory linear relationship between oxidase activity and copper concentration. A few more critical studies have indicated this relationship. The ceruloplasmin (PPD oxidase) assay procedure described herein has been so evaluated, and a very strong positive linear correlation between the total copper content of human serum and the PPD oxidase activity exists (3). Since the determination of ceruloplasmin is easier to perform than the determination of copper it may often be a more preferred and useful biochemical test than that of serum copper.

Principle of Assay

The enzymatic assay of ceruloplasmin activity is conveniently accomplished according to the procedure of Houchin (4) by measuring spectrophotometrically the color intensity produced when the serum sample is incubated with a buffered solution (pH 5.2) of PPD. Rice modified the acetate buffer by adding a small quantity of EDTA to suppress nonenzymatic oxidation of the substrate which is mainly due to trace-contamination of cupric ions. The enzymatic reaction is terminated by adding sodium azide solution prior to spectrometry. The results may be expressed in terms of absorbance, or more ideally, as "International Enzyme Units" (5, 6). The details for the standardization of ceruloplasmin activity in terms of the latter unitage are presented following the description of the assay procedure.

Reagents

1. *Acetate buffer, pH 5.2, ionic strength 1.2.* This solvent for the PPD contains 20.0 ml. glacial acetic acid, 163.0 g. sodium acetate trihydrate, and 15 mg. disodium dihydrogen ethylenediaminetetraacetate (EDTA) per liter of solution. The solution is stable indefinitely at room temperature.

2. *Buffered substrate solution.* This solution is freshly prepared 0.10% p-phenylenediamine (PPD) in acetate buffer.

NOTE: The substrate solution is stable for 24 hours when stored in the freezing compartment of a refrigerator. Later work has shown that a substrate solution using an equivalent amount (0.17%) of the dihydrochloride of PPD is more stable, and is therefore preferable. The commercial PPD · 2 HCl (Eastman Kodak) salt is purified by dissolving it in a minimum volume of hot (60°C.) water, adding Darco charcoal, and filtering while hot. The purified salt is precipitated from the filtrate by adding acetone until turbidity appears, refrigerating several hours, filtering off the crystals, and drying in the dark in a vacuum desiccator over anhydrous calcium chloride. The white crystals of purified PPD · 2 HCl are stored in a brown bottle in a refrigerator.

3. *Sodium Azide, 0.02%.* The solution is stable indefinitely at room temperature.

Procedure of Assay

Pipet 1.0 ml. of buffered substrate solution into each of two 16 × 100 mm. test tubes. Label one tube "test" and the other "control." Place both tubes in a 37°C. water bath and allow to come to bath temperature. To the "test" tube add 0.10 ml. of serum (or heparinized plasma), mix, and allow to incubate for exactly 15 minutes after the addition of the sample.

NOTE: Serum may be stored at 4°C. for several days, or frozen for several weeks without alteration of the ceruloplasmin activity.

At the end of the incubation period add 5.0 ml. of sodium azide solution to both the "test" and the "control" tubes. To the "control" tube add 0.10 ml. of serum and mix the contents of both tubes thoroughly. Determine the absorbance of the "test" against the "control" at a wavelength of 540 mμ.

NOTE: A Beckman Model B spectrophotometer with 1-cm. Pyrex absorption cells is used routinely by the submitter, but other instruments may be employed. The absorbance of the purple solution may be measured promptly and its color is stable for at least several hours.

If the absorbance of the "test" is greater than 0.2, repeat the analysis using serum diluted with an equal volume of saline.

NOTE: This dilution improves the correlation between the oxidase activity and the copper values with hypercupremic sera (3).

Expression of Results

Serum ceruloplasmin activity is often expressed in terms of arbitrary units, based on the increase in absorbance per unit time under the conditions of the particular procedure used. Thus, in the present case results may be recorded simply as the absorbance at 540 mμ attained by the "test" during the 15-minute incubation period (3).

NOTE: Several artificial reference standards for establishing ceruloplasmin enzyme unitage have also been proposed, including ferric salicylate (7), an acidic solution of the dye Pontacyl violet 6 R (8), and solutions of ceruloplasmin crystallized from pig serum (9). Neither of the former two substances has an absorption spectrum identical to that of PPD oxidized by ceruloplasmin (6), and the use of pig ceruloplasmin is not to be recommended in view of the observed differences in PPD oxidase activity in the sera of various mammals (10).

Whenever practicable, clinical enzyme units should be defined in International Enzyme Units, i.e., as micromoles of substrate transformed or of product formed per minute under specified conditions, and their concentration in terms of a milliliter (or liter) of serum, plasma, or urine (5). Previous to the technique described forthwith (6) such an ideal standardization was impossible in the case of ceruloplasmin because of the uncertainty as to the nature of the pigment formed by the enzymatic oxidation of PPD by serum.

Principle of Standardization

Upon oxidation of PPD in aqueous ammonical solution with hydrogen peroxide, a pure crystalline oxidation product forms slowly and is readily obtained by filtering the solution. This compound was first synthesized by Bandrowski in 1889 and has retained the name of "Bandrowski's base." It has an empirical formula of $C_{18}H_{18}N_6$, a molecular weight of 318.4, and has been assigned the following structural formula by Lauer and Sunde (11, 12).

It has been found that the pigment formed in the PPD assay procedure for ceruloplasmin has an absorption spectrum identical to that of Bandrowski's base with an absorption maximum at 540 mμ (6). The standardization of ceruloplasmin activity in terms of International Enzyme Units is thereupon made feasible by the use of standard solutions of Bandrowski's base.

Additional Reagents

1. Bandrowski's base[1] (13). The base is synthesized by dissolving 10 g. of PPD (recrystallized once from benzene) in 750 ml. of water containing 3 ml. of ammonium hydroxide and treating the solution with 125 ml. of 3% hydrogen peroxide. After standing for 24 hours in the dark, the solution is filtered through a Buchner funnel, the crystalline product is washed with water, and dried to a constant weight in a vacuum desiccator which is kept in the dark. The crystals melt sharply at 238–240°C., which agrees with data recorded in the literature.

2. Standard solutions of Bandrowski's base. A stock solution of the base is prepared using the acetate buffer, pH 5.2, as solvent. The solution is prepared by transferring an accurately weighed quantity of approximately 5 mg. of the dried crystals to a dry 100 ml. volumetric flask made of "Low Actinic" glass. About 75 ml. of acetate buffer is added in such a way that any particles adhering to the neck of the flask are washed into it.

NOTE: If this is not done all of the solid may not go into solution during the ensuing process.

The flask is shaken for 2 hours on a mechanical shaker, the contents diluted to the mark with buffer and the well-mixed solution (containing approximately 50 μg. of Bandrowski's base per ml.) used within an hour to establish a standard curve by the procedure which follows. After this period of time the solutions begin to fade slowly. Additional standard solutions containing approximately 5, 10, and 25 μg. of the dye per milliliter are prepared by diluting portions of the stock solution appropriately with acetate buffer.

Procedure of Standardization

The simple standardization technique is designed to duplicate the conditions of the ceruloplasmin assay procedure. Pipet 1.0 ml. of each of the standard solutions into four 16 × 100 mm. test tubes, each containing 5.0 ml. sodium azide solution and 0.10 ml. of serum, and mix thoroughly.

NOTE: Any unhemolyzed, nonlipemic, nonicteric serum may be used.

[1] Available from Nutritional Biochemicals Corp., 21010 Miles Ave., Cleveland, Ohio.

Determine the absorbances at a wavelength of 540 mμ against a "reagent blank" which contains sodium azide, serum, and acetate buffer (without base) in the amounts noted above. The color is stable for 30 minutes.

The absorbance (A) measurements at 540 mμ adhere strictly to Beer's Law up to concentrations of 50 μg. per ml. of standard solution. The mean absorptivity, a, defined as A divided by the product of the concentration of the base (in gm./l.) and the sample path length (in cm.), of three different preparations of base subjected to the prescribed procedure equaled 6.0. Using this absorptivity the molar absorptivity, ϵ, of the base, i.e., the product of a and the molecular weight of 318.4, is 1910.

Calculations

Assuming this value of ϵ, the absorbance (A) of a serum ceruloplasmin assay divided by 1.910 would equal the number of micromoles of Bandrowski's base formed per 15 minutes per 0.1 ml. of serum. It is desired to express ultimately micromoles of Bandrowski's base formed per minute per liter of serum under the specified conditions. Hence, the following calculation gives a final multiplication factor of 349: serum ceruloplasmin activity, in

$$\text{International Units} = \frac{A}{1.91} \times \frac{10,000}{15} = A \times 349$$

NOTE: The principle of this standardization technique using standard solutions of Bandrowski's base is applicable to other published procedures for the assay of ceruloplasmin which are based on the enzymatic oxidation of PPD (7, 14, 15). Naturally, different factors will apply depending on the actual method and on the spectrometric instrumentation employed. The modified Houchin procedure was chosen by the submitter because it had been previously evaluated (3).

Range of Normal Values

The accepted normal range of total serum copper in adults is 75–140 μg. per 100 ml. with a mean of approximately 105 μg. (3). These concentrations correspond to calculated ceruloplasmin absorbances of 0.102, 0.185, and 0.140, respectively (3), which, in terms of the International Units herein established, are 35.6, 64.6, and 48.8. Thus, for clinical purposes, the normal range of serum ceruloplasmin activity in adults may be considered to be 35–65 with an average of about 50 International Units.

Pathological Values

Several recent reviews have summarized the present knowledge of copper metabolism in humans (1, 16, 17). A great many pathological conditions are accompanied by a marked increase in plasma copper and ceruloplasmin. Virtually nothing of the significance of these increases is understood. Their chief clinical importance to date lies in the conclusion that serum copper (ceruloplasmin) plays the role of an "acute-phase reactant" (3, 18, 19). The increase in the ceruloplasmin activity that becomes evident at about the eighth week of normal pregnancy is also a consistent finding. Decreases in the plasma concentration of ceruloplasmin may result from the abnormally low absorption of copper found in tropical and nontropical sprue and in scleroderma of the small bowel, and in generalized abnormalities in protein metabolism such as those found in some infants with hypoproteinemia, hypoferremia, and anemia, in kwashiorkor, in the renal protein losses of nephrosis, or the intestinal protein losses of hypercatabolic hypoproteinemia or protein-losing enteropathy. In none of these conditions are there any ill effects that can be ascribed to ceruloplasmin deficiency and, in all of them, the deficiency of ceruloplasmin results from the deficiency in the copper and/or protein moieties of ceruloplasmin that the disease has brought about. Heriditary deficiency of ceruloplasmin (Wilson's disease) contrasts sharply to the foregoing since fatal pathologic changes are ultimately associated with this inherited abnormality. Although almost every patient with Wilson's disease has a deficiency or absence of ceruloplasmin, he has an excess of copper in virtually all other tissues. Early in life these deposits are unassociated with pathologic changes but eventually, severe histologic and functional damage appears, predominantly in liver and brain.

REFERENCES

1. Scheinberg, I. H., and Sternlieb, I., Copper metabolism. *Pharmacol. Rev.* **12**, 355–381 (1960).
2. Hirschman, S. Z., Morell, A. G., and Scheinberg, I. H., The heterogeneity of the copper-containing protein of human plasma, ceruloplasmin. *Ann. N.Y. Acad. Sci.* **94**, 960–969 (1961).
3. Rice, E. W., Correlation between serum copper, ceruloplasmin activity and C-reactive protein. *Clin. Chim. Acta* **5**, 632–636 (1960).
4. Houchin, O. B., A rapid colorimetric method for the quantitative determination of copper oxidase activity (ceruloplasmin). *Clin. Chem.* **4**, 519–523 (1958).

5. King, E. J., and Campbell, D. M., International enzyme units. An attempt at international agreement. *Clin. Chim. Acta* **6**, 301–306 (1961).

6. Rice, E. W., Standardization of ceruloplasmin activity in terms of International Enzyme Units. Oxidative formation of "Bandrowski's base" from *p*-phenylenediamine by ceruloplasmin. *Anal. Biochem.* **3**, 452–456 (1962).

7. Henry, R. J., Chiamori, N., Jacobs, S. L., and Segalove, M., Determination of ceruloplasmin oxidase in serum. *Proc. Soc. Exptl. Biol. Med.* **104**, 620–624 (1960).

8. Goldenberg, H., and White, D. L., Standardized method for the assay of serum oxidase activity (ceruloplasmin). *Clin. Chem.* **4**, 551 (1958).

9. Schimizu, M., Maruyama, Y., Kukita, M., Yanagisawa, Y., Sato, T., and Osaki, S. On the determination of serum ceruloplasmin and the results of its measurement. *J. Biochem.* **49**, 673–681 (1961).

10. McCoskor, P. J., Paraphenylenediamine oxidase activity and copper levels in mammalian plasmas. *Nature* **190**, 887–889 (1961).

11. Lauer, W. M., and Sunde, C. J., Structure and mechanism of the formation of the Bandrowski's base. *J. Org. Chem.* **3**, 261–264 (1938).

12. Sunde, C. J., and Lauer, W. M., Structure of the Bandrowski base. II N,N'-bis (2,5-diaminophenyl)-*p*-quinonediimine. *J. Org. Chem.* **17**, 609–612 (1952).

13. Ritter, J. J. and Schmitz, G. H., The constitution of Bandrowski's base. *J. Am. Chem. Soc.* **51**, 1587–1589 (1929).

14. O'Reilly, S., Observations on ceruloplasmin and methods for its estimation. *Neurology* **11**, 259–265 (1961).

15. Ravin, H. A., An improved colorimetric enzymatic assay of ceruloplasmin. *J. Lab. Clin. Med.* **58**, 161–168 (1961).

16. Sternlieb, I., and Scheinberg, I. H., Ceruloplasmin in health and disease. *Ann. N. Y. Acad. Sci.* **94**, Art. 1, 71–76 (1961).

17. Adelstein, S. J. and Vallee, B. L., Copper metabolism in man. *New Engl. J. Med.* **265**, 892–897, 941–946 (1961).

18. Rice, E. W., A study on correlations between C-reactive protein and certain other acute-phase reactants. *Clin. Chim. Acta* **6**, 170–173 (1961).

19. Rice, E. W., Evaluation of the role of ceruloplasmin as an acute-phase reactant. *Clin. Chim. Acta* **6**, 652–655 (1961).

SERUM CHOLINESTERASE*

Submitted by: HOWARD J. WETSTONE and GEORGE N. BOWERS, JR., Departments of Medicine and Pathology, Hartford Hospital, Hartford, Connecticut

Checked by: NELSON T. YOUNG, Medical College of Virginia, Richmond, Virginia
YASUO TAKENAKA, Children's Hospital, Honolulu, Hawaii

Introduction

This colorimetric method based on the hydroxylamine reaction as studied by Hestrin (2) has made it possible to perform this valuable clinical test with reproducibility, technical simplicity, and economy of time. Much of the prior confusion about serum cholinesterase metabolism stemmed from the inability to compare results obtained by a variety of analytical methods. In general, unskilled personnel may find measurements of enzyme activity by manometric, titrimetric, or potentiometric techniques less reproducible, more cumbersome, and less economical than the simple colorimetric method now available (3, 4, 5, 6, 7, 8). Other excellent colorimetric methods include those depending on aromatic substrates and phenol red (9) changes due to released acid during substrate hydrolysis.

In most of the quantitative methods, the enzymatic hydrolysis of an organic ester of choline under controlled conditions of sample size, substrate concentration, time, temperature, pH, ionic strength, and various salt concentrations results in the liberation of a choline salt and an organic acid. One of the reaction products or the residual substrate concentration is then measured with cholinesterase activity expressed as micromoles of substrate consumed per hour per milliliter of serum.

Principle

This procedure uses 0.2 ml. of serum as the source of enzyme. One hundred micromoles of acetylcholine bromide is employed as substrate. A maximum zero-order reaction rate is assured at this substrate

---

* Based on the method of de la Huerga (1) and Hestrin (2).

concentration (45 mM) by barbital buffer at pH 8.3 ± 0.1 and optimal ionic strength (about 0.2 M). After 60 minutes incubation of the enzyme-substrate mixture, alkaline hydroxylamine is added. This inactivates the protein and reacts stoichiometrically with the residual acetylcholine to form acethydroxamic acid as follows:

$CH_3COOCH_2CH_2N(CH_3)_3Br + NH_2OH \rightarrow CH_3CONHOH +$

$$HOCH_2CH_2N(CH_3)_3Br$$

The pH is adjusted to 1.2 ± 0.2 with hydrochloric acid. At this acidity a red-purple color which remains stable for at least 4 hours is fully developed within 10 minutes of the addition of ferric chloride.

$$CH_3CONHOH + FeCl_3 \longrightarrow CH_3C \underset{NHO}{\overset{O \dashrightarrow Fe^{+++}/3}{\diagup}}$$

Reagents

1. Barbital buffer, pH 8.3 ± 0.1. Dissolve 10.3 g. of sodium barbital in 300 ml. of water.[1] Add 22 ml. of 1 N HCl slowly and mix. Make up to 500 ml. with water. This buffer is stable for 2 months at 4°C.

2. Acetylcholine bromide, 0.50 M. Since acetylcholine bromide is extremely hygroscopic the following weighing procedure achieves maximum uniformity in substrate concentration. A 50-ml. beaker containing 25 ml. of water is weighed. Approximately 12 g. (but not less than 11.3 g.) of acetylcholine bromide (Eastman) are rapidly added to the water and the weight of the added substrate determined to the nearest milligram. The following formula is then used to calculate the final volume necessary to give a 0.5 M substrate solution:

$$\frac{\text{Weight of added acetylcholine bromide}}{11.306} = \frac{\text{final volume}}{100 \text{ ml.}}$$

The weighed substrate solution along with two washings is transferred to a 100-ml. flask and diluted to the mark with water. The additional water needed to give the calculated final volume is delivered from a pipet (1 ml. = 500 μmoles of substrate).

3. Substrate-buffer. (Must be prepared just prior to use.) One volume of acetylcholine bromide solution plus 9 volumes of barbital buffer (1 ml. = 50 μmoles of substrate).

---

[1] All water is distilled and deionized.

4. *Hydroxylamine HCl solution, 14%.* Dissolve 14 g. per 100 ml. of water. This solution is stable for 1 week at 4°C.

5. *NaOH, 14% (3.5 N).* Dissolve 14 g. per 100 ml. of water.

6. *Alkaline hydroxylamine solution.* Equal volumes of reagents (4) and (5) are mixed just prior to use. This solution is stable for 3 hours at room temperature.

7. *HCl, 0.6 N.* Dilute 50 ml. of concentrated HCl to 1000 ml.

8. $FeCl_3 \cdot H_2O$, *1%.* Dissolve 10 g. per 1000 ml. of 0.02 N HCl. Stable at room temperature. (It is convenient to keep a stock solution 10 times more concentrated.)

Procedure

1. Pipet 0.20 ml. of serum into a 15 × 125 mm. test tube. The daily 100 μmole control and standards are prepared by substituting 0.20 ml. of water for the serum in the analysis.

2. Add 2.0 ml. (100 μmoles of substrate) of substrate-buffer to consecutive tubes at 30-second intervals.

3. After 60 minutes incubation at 37°C. add 2.00 ml. of alkaline hydroxylamine solution and adjust the pH by addition of 6 ml. of 0.6 N HCl solution after a mininum waiting period of 1 minute.

4. Mix thoroughly by inversion of tubes at least 5 times.

5. Transfer 0.50 ml. of this mixture to another 15 × 125 mm. test tube; add 10.0 ml. of 1% ferric chloride and mix by inversion.

6. After a minimum waiting period of 10 minutes read the absorbance at 495 mμ (495 ± 10 mμ). The blank contains 0.50 ml. of HCl in 10 ml. of $FeCl_3$. The absorbance of the blank is set to zero. (Bubbles, should they appear, are released by gentle tapping of the cuvette prior to reading.)

7. Cholinesterase activity of the sample is obtained by reference to a calibration curve of standards which have been treated (see Fig. 1) the same as the serum sample (including incubation at 37°C.).

Calibration Curve of Standards

It is suggested that the calibration curve be based on several paired sample determinations at various substrate concentrations within the range of concentrations to be measured. Repeating this procedure for each new batch of acetylcholine bromide over a period of a few months will give a fairly constant curve. It is then suggested that points on the curve be checked at frequent intervals and that the

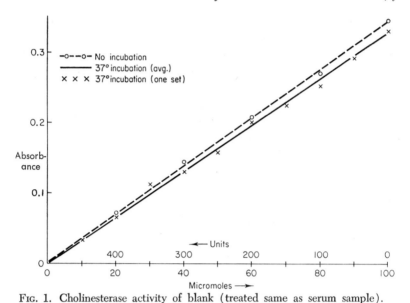

FIG. 1. Cholinesterase activity of blank (treated same as serum sample).

absorbance of a 100 μmole control be measured daily to detect possible substrate contamination. The minor daily variations in the control absorbance should not exceed ±0.02.

The following procedure illustrates the method for determining points on the curve:

1. Prepare the standard substrate-buffer as before (Solution A).

2. Prepare Solution B as follows: 1 volume water and 9 volumes buffer.

3. The accompanying tabulation indicates the volume of Solution A and Solution B to be added to 0.2 ml. of water (making a total volume of 2.2 ml.) to achieve various substrate concentrations.

Substrate concentration (μmoles)	Solution A (ml.)	Solution B (ml.)
100	2.0	None
90	1.8	0.2
80	1.6	0.4
70	1.4	0.6
60	1.2	0.8
50	1.0	1.0
. . .	. . .	. . .
10	0.2	1.8
0	None	2.0

4. Continue the test procedure at Step 3.

The cholinesterase activity is related to the difference between the 100 μmoles of acetylcholine added and the micromoles remaining after incubation with the sample (in other words the substrate consumption) multiplied by 5 (to correct to 1 ml. of serum). The serum enzyme activity is read directly from the chart (see Fig. 1, solid line). An alternate method involves calculating the result by the following formula:

$$\frac{\text{A of } 100\mu\text{moles of incubated standard} - \text{A of sample}}{\text{A of } 100 \ \mu\text{moles of nonincubated standard}} \times 500 = \begin{array}{c} \text{cholinesterase} \\ \text{activity} \end{array}$$

Comments

Cholinesterase activity should be measured in fasting venous blood drawn in a manner to avoid hemolysis. Serum and plasma values are similar. The samples are stable for several days at 4°C. and for several weeks if frozen.

Although it has been stated that the amount of acetylcholine bromide used in this procedure is sufficient to saturate the enzyme throughout the entire incubation (1) it has been shown (10) that with higher enzyme activities there is a slowing of the reaction rate during the latter part of a 60 minute incubation period. Since, unlike acetylcholinesterase, serum cholinesterase is not inhibited by an excess of substrate it is suggested that a sample of 0.1 ml. of serum be used where higher values are suspected or found by a prior enzyme determination. For practical clinical purposes this is only of importance in patients with essential hypertension or nephrosis. Under certain circumstances there may be situations where smaller samples would be used for all determinations.

The procedure described here has been found to be most easily reproduced and simple.

The color produced by the ferric-acet-hydroxamic acid complex shows a maximum absorption at 495 mμ against a ferric chloride blank on the Beckman DU instrument (Fig. 2). Since the peak is relatively flat any photometer capable of measurements near 500 mμ is satisfactory. The color complex shows conformity to Beer's Law with no interference from the hydrolysis products (acetate and choline bromide). A slight nonenzymatic hydrolysis occurs during the incubation and for this reason the standards used in preparation of the calibration curve are also incubated prior to color development and reading.

Technicians with considerable experience in this procedure can

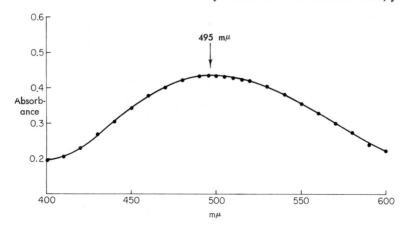

Fig. 2. Absorption of color produced by FeCl₃ hydroxamic acid with 100 μM acetylcholine–BR vs. blank.

readily achieve a precision (based on paired sample determinations) of between 5 and 8% (95% of determinations) (10). Results of a daily serum pool checked by eight technicians rotating at 2-week intervals indicated a maximum error of 8% in 95% of determinations.

Normal and Pathological Values

(See references 10, 11, 12, 13, 14, 15, 16. Reference 15 contains an extensive bibliography.)

Table I outlines values for serum cholinesterase in a variety of non-

TABLE I

SERUM CHOLINESTERASE IN A VARIETY OF NORMAL SITUATIONS

Classification	No.	Mean	σ	Observed range	per cent less than 150 units
Normal adult: fasting	51	204	36.0	155–288	0
Normal adult: postprandial	64	220	38.5	150–305	0
Normal pregnancy					
First trimester	17	175	35.2	90–245	17.6
Second trimester	70	159	39.4	35–250	32.9
Third trimester	95	148	38.3	35–238	43.2
During Labor	61	145	40.2	48–258	47.5
Fetal cord blood	61	130	40.2	29–253	72.1
6 Weeks postpartum	43	183	35.1	118–278	7.0

disease situations. Normal adults have significantly higher values postprandially than in the fasting state. Mean values fall as pregnancy progresses and newborn children have a quite low mean enzyme activity. Table II indicates the abnormality of the enzyme in a series of patients with liver disease. Table III compares results in patients

TABLE II
SERUM CHOLINESTERASE IN LIVER DISEASE (ADULTS)

Classification	No.	Mean	Per cent less than 150 units	Per cent less than 205 units
Viral hepatitis	31	151	42	90
Compensated cirrhosis	22	171	36	77
Decompensated cirrhosis	40	87	88	95
Obstructive jaundice: benign	25	179	36	60
Obstructive jaundice: malignant	54	108	82	94

TABLE III
SERUM CHOLINESTERASE IN CARCINOMA (ADULTS)

Classification	No.	Mean	Per cent less than 150 units
Benign disease: out patients[a]	102	205	5
Benign disease: in patients[a]	103	178	29
Carcinoma	237	124	75
Local carcinoma[b]	41	157	56
Nonhepatic metastatic[b]	93	136	63
Hepatic metastatic[b]	74	92	85

[a] Nonhepatic, nonhypertensive.
[b] Based on a series of 208 patients evaluated for the purpose of extent of involvement by carcinoma.

with benign and malignant disease. Patients with hypertension have a higher mean value (254 units) than normal adults.

In all circumstances correlation coefficients indicate that serum albumin and serum cholinesterase are not measures of the same or a similar function but are in large degree independent parameters of liver function and protein metabolism (16).

Clinical Usefulness

Enzymes such as the transaminases, dehydrogenases, aldolases, and certain peptidases have considerable value when they appear in elevated amounts in serum. Such elevations in activity, however, seem to be a result (direct or indirect) of the acute injury of normal or pathologic tissue and do not reflect the total residual function of the organ in question. On the other hand serum cholinesterase is depressed in liver disease due to a decreased synthetic capacity. The depression in carcinoma is partially due to the tumor itself. Perhaps this is an effect of acting through an enzyme inhibitor from the carcinoma (16, 17).

Although serum and red blood cell cholinesterase metabolism are still under investigation in a host of clinical and research situations measurement is of proven value in the following:

A. In Hepatic Disease (Nonpregnant Adults)

1. *Viral Heptatitis.* Although the initial determination may be within the normal range it is almost always below the mean value for normals. Much more important than its use as a diagnostic aid is the information obtained by serial determinations which show a steady rise in activity in the patient who is recovering, a flat response in the smoldering case, and a sharp decline with increasing severity of disease (often in advance of clinical change).

2. *Cirrhosis.* The level of cholinesterase activity is a reliable index of the over-all involvement of the liver and often correlates more closely with the clinical and morphologic findings than do other standard liver function tests.

3. *Obstructive Jaundice.* Over three-quarters (75 cases) of 94 patients with obstructive jaundice could be diagnosed with 88% accuracy as to malignant or benign etiology (as determined by surgical exploration) by their serum cholinesterase activity. The remaining one-quarter were not diagnosable by this means which utilized values below 150 as indicative of tumor and those over 185 of benign disease. Patients with values below 150 units almost always have malignant disease as the cause of their obstruction whereas those with enzyme activity over 185 units usually have a benign etiology.

B. In Carcinoma

Values below 150 units (in nonpregnant adults who do not suffer from primary liver disease) in general indicate the possibility of tumor being present. Over 80% (284 cases) of 340 patients in the hospital with either benign nonhepatic disease or carcinoma could be diagnosed with 81% accuracy by their serum cholinesterase activity. Although it is clear that this enzyme test is not by any means a cancer diagnostic test it may be useful as a screening device in the clinical evaluation of individual patients. In addition an adult with a previously confirmed neoplasm who shows a decline in enzyme activity is suspect of spreading or recurrence of disease.

C. Poisoning (Ed.)

Most hospitals use cholinesterase measurements to detect poisoning by organic phosphate insecticides or some drugs. Marked depressions are associated with respiratory failure.

References

1. de la Huerga, J., Yesinick, C., and Popper, H., Colorimetric method for the determination of serum cholinesterase. *Am. J. Clin. Pathol.* **22**, 1126 (1952).
2. Hestrin, S., The reaction of acetylcholine and other carboxylic acid derivatives with hydroxylamine and its analytical application. *J. Biol. Chem.* **180**, 249 (1949).
3. Alles, G. A., and Hawes, R. C., Cholinesterases in the blood of man. *J. Biol. Chem.* **133**, 375 (1940).
4. Ammon, R., Die fermentative Spalung des Acetylcholine. *Arch. Ges. Physiol.* **233**, 486 (1933).
5. Augustinsson, K.-B., Cholinesterases: study in comparative enzymology. *Acta Physiol. Scand. Supp.* 52 **15**, 1 (1948).
6. Glick, D., Properties of choline esterase in human serum. *Biochem. J.* **31**, 521 (1937).
7. Glick, D., Studies on enzymatic histochemistry. XXV. A micromethod for the determination of choline esterase and the activity-pH relationships of the enzyme. *J. Gen. Physiol.* **21**, 289 (1938).
8. Michel, H. O., An electrometric method for the determination of red blood cell and plasma cholinesterase activity. *J. Lab. Clin. Med.* **34**, 1564 (1949).
9. Reinhold, J. G., Tourigny, L. G., and Yonan, V. L., Measurement of serum cholinesterase activity. By a photometric indicator method. Together with a study of the influence of sex and race. *Am. J. Clin. Pathol.* **23**, 645 (1953).
10. Wetstone, H. J., Tennant, R., and White, B. V., Studies of cholinesterase activity. I. Serum cholinesterase, methods and normal values. *Gastroenterology* **33**, 41 (1957).
11. LaMotta, R. V., Williams, H. M., and Wetstone, H. J., Studies of cholinesterase activity. II. Serum cholinesterase in hepatitis and cirrhosis. *Gastroenterology* **33**, 50 (1957).
12. Williams, H. M., LaMotta, R. V., and Wetstone, H. J., Studies of

cholinesterase activity. III. Serum cholinesterase in obstructive jaundice and neoplastic disease. *Gastroenterology* **33,** 58 (1957).
13. White, B. V., Wetstone, H. J., and LaMotta, R. V., Serum cholinesterase activity in malignant neoplasms. *Trans. Am. Clin. and Climat. Ass'n.* **69,** 176 (1957).
14. Wetstone, H. J., LaMotta, R. V., Middlebrook, L., Tennant, R., and White, B. V., Studies of cholinesterase activity. IV. Liver function in pregnancy, Values of certain standard liver function tests in normal pregnancy. *Am. J. Obstet. Gynecol.* **76,** 480 (1958).
15. Reinfrank, R., and Wetstone, H. J., Serum cholinesterase activity in hypertension. Report of a pilot study. *Htfd. Hosp. Bull.* **13,** 32 (1958).
16. Wetstone, H. J., LaMotta, R. V., Bellucci, A., Tennant, R., and White, B. V., Studies of cholinesterase activity. V. Serum cholinesterase in patients with carcinoma. *Ann. Internal Med.* **52,** 102 (1960).
17. LaMotta, R. V., Babbott, D., and Wetstone, H. J., Studies of cholinesterase activity. VI. The depression of serum cholinesterase activity by a locally implanted human tumor in the guinea pig. *Cancer Res.* **21,** 749 (1961).

Addendum

Serum cholinesterase is listed by the Commission on Enzymes of the International Union of Biochemistry (1961) as acylcholine acyl-hydrolase (3.1.1.8). The international unit is 1/60 the conventional clinical unit referred to in this chapter.

COPPER IN SERUM*

Submitted by: EUGENE W. RICE, W. H. Singer Memorial Research Laboratory of
the Allegheny General Hospital, Pittsburgh, Pennsylvania

Checked by: EDWARD MONTONI, Franklin County Public Hospital, Greenfield,
Massachusetts
EDNA ANDREWS, Metropolitan Hospital, Detroit, Michigan

Introduction and Principles

In 1956 Gran (1) described the use of oxalyldihydrazide for the sensitive spectrophotometric microdetermination of copper. This reagent, in the presence of formaldehyde or acetaldehyde in ammoniacal solution, produces an intense lavender color with copper (II). The molar absorbance index of the resulting complex averages 22,000 at 542 mμ, in contrast to indices of 16,000, 14,000, and 8,000 for the copper-complexes with cuprizone, bathocuprione, and diethyl dithiocarbamate, respectively. Gran found that the presence of reasonable amounts of hydrochloric, nitric, and perchloric acids or of sodium sulfate had no influence on absorbance.

Stark and Dawson (2) have further evaluated the reaction and applied it to the determination of copper in purified preparations of copper oxidases. They reported that their procedure determined as little as 0.1 μg. of copper per ml. in the presence of protein with a precision of 2 to 3%. Manganese (II) interfered somewhat but cadmium (II), nickel (II), magnesium (II), iron (II), zinc (II), cobalt (II), and calcium (II) did not. Gelatin, β-lactoglobulin, and glutathione did not interfere, but bovine plasma albumin did, and this interference was quantitative.

The following describes the application of oxalyldihydrazide to a rapid, accurate, and precise procedure for the spectrophotometric determination of serum total copper (3). The serum is treated with a hydrochloric acid solution of oxalyldihydrazide which extracts the copper from its combination with serum proteins (4) and also introduces the reagent into the system. A trichloroacetic acid filtrate of the mixture is subsequently prepared. Ammonium hydroxide and acetaldehyde

---

* Based on the method of Rice (3).

57

58

are added to an aliquot of the filtrate, and the absorbance of the resulting stable lavender color is determined at 542 mμ, the wavelength of maximum absorbancy of the colored complex.

Reagents

It is imperative that all glassware be washed with soap and water, soaked overnight in 1:1 nitric acid followed by rinsing with hot tap water, and finally with generous portions of distilled-deionized water. Such glassware should be reserved for trace metal determinations only and must be protected carefully from contamination after it has been cleaned. Resin deionized-distilled water is employed for all reagents instead of the conventional "glass double-distilled" water.

1. Oxalyldihydrazide. Dissolve 146 g. of diethyl oxalate in 770 ml. of absolute ethanol. Also, dissolve 59 g. of hydrazine hydrate (85% solution, purified) in 295 ml. of absolute ethanol and mix the two solutions. In less than an hour a nearly quantitative yield of white precipitate forms. Recrystallize the crude product from boiling water (m.p. 239°C. with decomposition). Oxalyldihydrazide is available commercially as Eastman catalogue no. 7157. Recrystallize this product once from boiling water.

2. Hydrochloric acid, 2.0 N, containing oxalyldihydrazide, 0.10%.

3. Trichloroacetic acid, 20.0%. Use sulfate-free, iron-free, highest purity chemical, Eastman catalogue no. 259.

4. Ammonium hydroxide, concentrated, sp.gr. 0.90.

5. Acetaldehyde aqueous solution, 50% (v/v). Prepare this reagent from 1 part cold water and 1 part cold acetaldehyde (reagent or technical grade). Store the reagent in the refrigerator.

NOTE: Reagents 2–5 are stable indefinitely.

6. (Ethylenedinitrilo) tetraacetic acid disodium salt, (EDTA).
7. Citric acid, crystal.

NOTE: Baker Analyzed reagents are satisfactory for Reagents 2, 4, 6, and 7.

8. Standard copper solution, 1.00 mg. Cu per ml. Dissolve 0.3928 g. of reagent grade cupric sulfate pentahydrate, $CuSO_4 \cdot 5H_2O$, in water, and dilute to exactly 100 ml. Prepare working standards daily by diluting this stock standard solution. Routinely, a standard solution containing 200 μg. Cu per 100 ml. is used. This is made by diluting the stock solution 1:500 with water.

Procedure

Into a 13 × 100 mm. test tube add 1.0 ml. serum or heparinized plasma and 0.70 ml. of 2.0 N HCl containing 0.10% oxalyldihydrazide. Mix, and let set for 10 minutes.

NOTE: In the submitter's laboratory blood is collected with "B-D Vacutainer" apparatus using "No. L-3200 tubes for blood lead determination" and new needles (Becton, Dickinson, and Co., Rutherford, N.J.). If syringes are used they must be cleaned as described above.

Add 1.0 ml. of 20.0% trichloroacetic acid, mix well with a thin stirring rod, cover the tube with Parafilm, let set for about 5 minutes, and centrifuge rapidly for 15 minutes. With each set of unknowns carry through the entire procedure 2 water blank tubes and 1 standard tube, (2.0 μg. Cu per ml.).

NOTE: Mixing with a "Vortex Mixer" is preferable to the use of a stirring rod.

Into test tubes containing a pinch of citric acid crystals, pipet 2.0 ml. of the clear supernatant fluid from each blank, standard, and unknown and agitate to dissolve the crystals.

NOTE: Citric acid is added to prevent possible interference from other metals, notably iron.

To one of the blanks add a pinch of EDTA. Do not add EDTA to the other blank.

NOTE: EDTA prevents formation of the color and hence, this defines its role in the reagent blank tube against which all other absorbancies are read. Because EDTA prevents the color development, it cannot be employed as an anticoagulant for obtaining plasma.

In all tubes mix 0.5 ml. ammonium hydroxide and 0.50 ml. cold acetaldehyde solution, in that order. Allow the tubes to remain at room temperature for at least 30 minutes, then determine the absorbances at 542 mμ against the blank containing the EDTA. The color is stable for at least several hours.

Calculation

$$\frac{(\text{Absorbance unknown}) - (\text{absorbance blank})}{(\text{Absorbance standard}) - (\text{absorbance blank})} \times 200 = \mu g. \ Cu/100 \ ml. \ serum$$

(With a Beckman B Spectrophotometer and 1-cm. Pyrex absorption cells, the net absorbance value of the divisor in this equation averages 0.185 ± 0.002).

Discussion

The lavender-complex conforms to Beer's Law at 542 mμ in concentrations of copper up to 5 μg. per 2 ml. of filtrate. This amount of copper is equivalent to a serum level of 675 μg. per 100 ml. The precision obtained by analyzing ten aliquots each of two pooled sera gave a mean standard deviation of 2.3 μg. Cu per 100 ml. Also, twenty-five sera analyzed in duplicate had a standard deviation of 2.7 μg. Cu per 100 ml., with an average difference between duplicates of 2.0 μg. As summarized in Table I, the average percentage recovery of various amounts of copper added to human sera is complete.

TABLE I

RECOVERIES OF ADDED AMOUNTS OF COPPER TO HUMAN SERA[a]

Serum	Cu found in original specimen	Cu added	Cu found after addition	Per cent recovery
1	184	10	193	99.5
2	155	20	169	96.6
3	90	30	117	97.5
4	133	40	177	102.2
5	172	40	205	103.4
6	121	50	171	100.0
7	166	60	218	96.5
8	195	80	281	102.1
9	146	80	221	102.1
10	145	100	247	100.8
11	146	100	249	101.1
12	121	140	258	98.8
			Mean	100.0

[a] Values represent μg. Cu per 100 ml.

Values obtained with the proposed method were compared with those obtained with the widely-used carbamate procedure of Gubler et al. (5). Typical data shown in Table II illustrate that, within experimental error, essentially identical results are obtained with both techniques. Gubler's method, however, is not as sensitive or as easy to perform as the present one.

Small variations in the amounts of reagents do not affect the color intensity (2). This procedure can also be applied to tissues or urine by employing a wet acid-digestion process. With this modification

TABLE II

COMPARISON OF THE PROPOSED AND THE GUBLER METHOD FOR
THE DETERMINATION OF SERUM TOTAL COPPER

	Cu (μg. per 100 ml.)		
Serum	Proposed Method	Gubler's Method	Percentage Deviation
1	140	134	+ 4.5
2	103	102	+ 1.0
3	136	121	+11.2
4	145	128	+11.2
5	143	136	+ 5.0
6	123	123	0.0
7	113	119	− 5.0
8	162	160	+ 1.2
9	133	143	− 7.0
10	134	130	+ 3.0
11	167	177	− 5.6
12	146	151	− 3.4
13	194	191	+ 1.5
14	183	173	+ 5.9
15	140	140	0.0
16	108	106	+ 1.9
17	145	155	− 6.4
18	149	151	− 1.2
19	184	179	+ 2.9
20	158	173	− 8.7
		Mean	+ 0.6

where hydrochloric acid (Reagent 2) is not used to release protein-bound copper, the oxalyldihydrazide is introduced into the system by saturating the concentrated ammonium hydroxide (Reagent 4) with this reagent. Such oxalyldihydrazide-ammonia solutions should be prepared daily since in contrast to Reagent 2, they slowly lose color-forming ability.

NOTE: Ed. 1.0 ml. urine + 0.5 ml. 70% perchloric acid digested to fumes can be processed as the TCA filtrate.

The submitter has also developed a micro procedure of this method by using 200 lambdas (λ) of serum and by reducing proportionately the volumes of reagents. Microliter (lambda) pipets and a Beckman Model B Spectrophotometer with "V Fused Silica" micro-absorption cells were used. The cells have a 10 mm. light path, are 25 mm. high, have a total capacity of 750 λ and require a special cell carrier.

NOTE: The microcells and the cell carrier are obtained from Pyrocell Manufacturing Co., 207 E. 84th Street, New York 28, N. Y.

Values obtained with the micro modification were compared with those obtained with the macro method by analyzing sixty sera by each technique. The difference between the two procedures ranged from −10 μg. to +15 μg. Cu per 100 ml. serum. The micro procedure gave a mean difference of only −0.4 μg. Cu per 100 ml., thereby showing that identical results may be expected with both techniques.

NOTE: Considerable effort was directed toward developing an ultramicro procedure using a system that is currently available commercially. These experiments have consistently given erratic results and it is the submitter's recommendation that scaling the technique down further than described would require considerable purification of reagents and careful evaluation by the analyst.

Subsequent to the original paper by Rice (3), Welshman (6) also proposed oxalyldihydrazide for a copper procedure using 1 ml. of serum and a final volume of 5 ml. and Summers (7) has published a similar method requiring 2 ml. of serum for a final volume of 3 ml. The latter investigator (7) moreover adapted an oxalyldihydrazide copper method to the AutoAnalyzer which needs "slightly more than 1.2 ml. of serum for each determination." Persons interested in a critical survey of the literature on the determination of trace amounts of copper in materials of biological origin should consult the review of Borchardt and Butler (8).

Range of Values

Studies of the copper content of serum in normal infants are few and the results lack agreement. Ranges of 80 to 180, 150 to 235, 93 to 244, and 83 to 284 μg. per 100 ml. have been suggested by different authors (9, 10). In contrast, similar studies in normal adults are sufficiently numerous and results are in such agreement that "normal ranges" can be more satisfactorily defined. The submitter's laboratory has confirmed in general the "normal" values established by Neal and Fischer-Williams (11). These are, for males, 76 to 133 μg. per 100 ml. (mean, 100; S.D. ±12) and for females, 70 to 140 (mean, 108; S.D. ±15). An excellent review by Scheinberg and Sternlieb (12) summarizes current knowledge of human copper metabolism. Hypercupremia occurs in a variety of conditions such as pregnancy, myocardial infarction, cirrhosis, rheumatoid arthritis, rheumatic heart disease, and most acute and chronic infections. Hypocupremia is associated with certain nutritional disorders, nephrosis, and Wilson's Disease. We understand virtually nothing of the physiologic significance of these changes. Their chief importance to date lies in the conclusion that serum copper, like the erythrocyte sedimentation rate or the leucocyte count, is an "acute-phase reactant" (12–14).

REFERENCES

1. Gran, G., The use of oxalyldihydrazide in a new Reagent for the spectrophotometric microdetermination of copper. *Anal. Chim. Acta* **14**, 150–152 (1956).
2. Stark, G. R., and Dawson, C. R., Spectrophotometric microdetermination of copper in copper oxidases using oxalyldihydrazide. *Anal. Chem.* **30**, 191–194 (1958).
3. Rice, E. W., Spectrophotometric determination of serum copper with oxalyldihydrazide. *J. Lab. Clin. Med.* **55**, 325–328 (1960).
4. Peterson, R. E., and Bollier, M. E., Spectrophotometric determination of serum copper with biscyclohexanoneoxalyldihydrazone. *Anal. Chem.* **27**, 1195–1197 (1955).
5. Gubler, C. J., Lahey, M. E., Cartwright, G. E., and Wintrobe, M. M., Studies on copper metabolism: Method for determination of copper in whole blood, red blood cells, and plasma. *J. Biol. Chem.* **196**, 209–220 (1952).
6. Welshman, S. G., The determination of serum copper. *Clin. Chim. Acta.* **5**, 497–498 (1960).
7. Summers, R. M., Microdetermination of serum copper. *Anal. Chem.* **32**, 1903–1904 (1960).
8. Borchardt, L. G., and Butler, J. P., Determination of trace amounts of copper. *Anal. Chem.* **29**, 414–419 (1957).
9. Schubert, W. K., and Lahey, M. E., Copper and protein depletion complicating hypoferric anemia of infancy. *Pediatrics* **24**, 710–733 (1959).
10. Bakwin, R. M., Mosbach, E. H., and Bakwin, H., Concentration of copper in serum of children with schizophrenia. *Pediatrics* **27**, 642–644 (1961).
11. Neal, F. C., and Fischer-Williams, M., Copper metabolism in normal adults and in clinically normal relatives of patients with Wilson's disease. *J. Clin. Pathol.* **11**, 441–447 (1958).
12. Scheinberg, I. H., and Sternlieb, I., Copper metabolism. *Pharmacol. Rev.* **12**, 355–381 (1960).
13. Rice, E. W., Correlation between serum copper, ceruloplasmin activity and C-reactive Protein. *Clin. Chim. Acta* **5**, 632–636 (1960).
14. Rice, E. W., A study on correlations between C-reactive protein and certain other acute-phase reactants. *Clin. Chim. Acta* **6**, 170–173 (1961).

URINARY ESTROGENS*

Submitted by: R. Hobkirk and Ann Metcalfe-Gibson, McGill University Medical Clinic and Department of Metabolism, The Montreal General Hospital, Montreal, Canada

Checked by: Nathan Kase, Yale University School of Medicine, New Haven, Connecticut
Edward Richardson, Delaware Hospital, Wilmington, Delaware

Introduction and Principles

In the past the chemical measurement of urinary estrogens has been a complex procedure unsuited to the routine laboratory. Within recent years, however, Ittrich (1, 2) has described simplified methods for the chemical assay of total estrogens as a group, in urine from either pregnant or nonpregnant individuals. These methods are based on a modification of the original Kober color reaction (3). The specific chromogen formed when urinary estrogen extracts are heated with aqueous H_2SO_4 in the presence of hydroquinone is extracted into a tetrachloroethane (or tetrabromoethane) solution of p-nitrophenol leaving considerable amounts of nonspecific chromogens in the aqueous phase. This results in a reaction of increased specificity and sensitivity by which the estrogen may be quantitatively estimated colorimetrically (if sufficient estrogen is present) or fluorimetrically (where very low estrogen concentrations are involved). In the pregnant state the color reaction is usually performed directly on the urine with no prior purification. In urines of the nonpregnant state and in sugar-containing pregnancy urines some purification is necessary prior to the application of the reaction. This involves preliminary hot acid hydrolysis of diluted urine to split estrogen conjugates; preferential removal of neutral contaminants with organic solvents; purification by alkaline treatment; pH adjustment and subsequent reextraction of estrogen with ether; fluorimetric measurement.

The earlier method of Bauld (4) for the separate measurement of the "classical" estrogens estrone, estradiol-17β, and estriol can be simplified for the estimation of estriol alone in pregnancy urine (5, 6). This approach has much to commend it since it allows the measure-

* Based on the methods of Ittrich (1, 2) and Bauld (4).

65

ment of a specific estrogen metabolite. During pregnancy urinary estriol levels are so high that relatively small volumes of urine, reagents, etc., can be employed in a method suited to the routine laboratory. The principles involved include hot acid hydrolysis; extraction of estrogens with ether and purification of the extracts by alkaline treatment; partition of the estrogen-containing fraction between benzene and water; treatment of the aqueous (estriol-containing) phase with alkali followed by pH adjustment with CO_2; reextraction of estriol with ether prior to measurement by a modified Kober color reaction (7).

Three routine chemical methods for urinary estrogen measurement are described in the following pages. These are (a) measurement of total estrogens in nonpregnancy urine; (b) measurement of total estrogens in pregnancy urine; (c) measurement of estriol in pregnancy urine.

Preparation of Patient and Urine Collection

Where possible withhold medication in the form of meprobamates, phenolphthalein, cascara, senna, stilbestrol, or large doses of cortisone during the period of urine collection and for 24 hours prior to it.

NOTE 1: All of these drugs may not interfere with the methods described here but each is known to be detrimental to certain estrogen methods (8). For this reason it is advisable to take such precautions.

Complete 24-hour urine collections are made without preservative in glass or polyethylene bottles. Analyses should be performed as soon as possible thereafter although urine may safely be stored at 4°C. for 1 week without apparent loss of estrogen or alternatively frozen for a longer period.

Cleaning of Glassware

All used urine bottles, test tubes, round-bottom flasks, and pipettes are rinsed with tap water, steeped in dichromate-sulfuric acid mixture overnight, washed with tap water, steeped overnight in ethanol, thoroughly rinsed with distilled water, and dried. Separatory funnels, Erlenmeyer flasks, and beakers are thoroughly washed with tap water and then rinsed with distilled water.

Ia. TOTAL ESTROGENS IN NONPREGNANCY URINE

Reagents

1. Hydrochloric acid, concentrated, reagent grade.

2. Sodium hydroxide, 10 N. Dissolve 40 g. of reagent grade NaOH pellets in about 60 ml. of distilled water (carefully with cooling) and make up to 100 ml. This solution need not be standardized. Store in a polyethylene bottle.

3. Benzene, reagent grade. Purify by shaking with several changes of concentrated H_2SO_4 (1/20 volume) until the acid layer is colorless or nearly so. Reject the final acid wash and shake the benzene with 1/20 volume of 0.1 N NaOH followed by 3 times with 1/10 volume of distilled water. After drying over anhydrous Na_2SO_4 distil twice in an all-glass apparatus rejecting suitable fractions (10–20%) at the beginning and end of the distillation. Store in a dark glass bottle at 4°C.

4. Hexane, reagent grade. Purify and store as for benzene.

5. Sodium hydroxide, 1 N. Dilute 10 ml. of 10 N NaOH to 100 ml. with distilled water. No standardization is required. Store in a polyethylene bottle.

6. Diethyl ether, reagent grade. Purify by shaking for approximately 3 minutes with 1/10 volume of 0.3 M $FeSO_4$ in 0.4 M H_2SO_4 (100 ml./l.) followed by brief shaking with 3 × 1/10 volume of distilled water. Distil once in an all-glass apparatus rejecting suitable fractions at beginning and end.

NOTE 2: This purification destroys ether peroxides which may be harmful to estrogens. The ether must be used within 6 hours of distillation.

7. Sodium bicarbonate solution, 8% (w/v). Dissolve 80 g. of reagent grade $NaHCO_3$ in distilled water at room temperature and dilute to 1 l.; filter and store in a polyethylene bottle.

8. Sodium carbonate solution, saturated. Saturate a volume of distilled water at 100°C. with reagent grade Na_2CO_3. Cool to room temperature (22°C.) and filter.

9. Sodium carbonate buffer, pH 10. Mix 88 ml. of 8% $NaHCO_3$ solution with 12 ml. of saturated Na_2CO_3. Check with short range test paper to ensure that the pH is 9.5–10. Adjust if necessary with either constituent solution, filter and store in a polyethylene bottle.

10. Ammonium sulfate solution, 8% (w/v). Dissolve 80 g. of reagent grade $(NH_4)_2SO_4$ in distilled water and dilute to 1 l. Filter if necessary.

11. *Ethanol, aldehyde and ketone-free.* Mix 2.5 g. of 2,4,dinitrophenylhydrazine and 0.5 ml. of concentrated HCl and 1 l. of 95% ethanol. Reflux at boiling point for 1 hour, distil and redistil using an all-glass apparatus. Reject suitable fractions at the beginning and end of distillations. Store the purified material in a dark glass bottle.

12. *Sulfuric acid, concentrated, reagent grade.*

13. *Hydroquinone, reagent grade.* Grind finely using a pestle and mortar.

14. *p-Nitrophenol in tetrabromoethane (or tetrachloroethane), 2% solution (w/v).* Dissolve 2.00 g. of p-nitrophenol (reagent grade twice crystallized from benzene) in 1.0 ml. of ethanol and dilute to 100 ml. with tetrabromo- or tetrachloroethane (reagent grade, previously dried with anhydrous Na_2SO_4). This solution is stable for at least 3 days if kept at 4°C.

NOTE 3: Tetrabromoethane results in a much more sensitive (twofold) fluorescent reaction than does the tetrachloro- compound. Also, the tetrabromo- compound apparently leads to a more specific reaction, extracting as it does less nonspecific background material (2). The tetrabromoethane is the choice solvent for fluorescent measurements. Turbidity which can occur with this solvent has little or no effect in the fluorescence measurement.

15. *Alkaline eosin solution, 10 µg./ml.* Dissolve 10 mg. of eosin (yellow) in 1 l. of carbonate buffer (reagent 9). Store in the refrigerator and allow to come to room temperature before using.

NOTE 4: This solution is employed as a stable source of fluorescence to set the scale during fluorimetry. The exact concentration required will depend on the type of instrument and filter combination used and also the range of estrogen concentration to be measured. The fluorimeter used should, if necessary, be equipped with a cooling device since the fluorescence of the eosin standard changes markedly with temperature.

Standard Estrogen Solutions

(A) Stock estriol solution, 1 mg./ml. Dissolve 100 mg. of pure crystalline estriol in purified ethanol and dilute to 100 ml. with ethanol.

(B) Dilute estriol solution, 10 µg./ml. Dilute 1.00 ml. of solution A to 100 ml. with ethanol.

(C) Dilute estriol solution, 2 µg./ml. Dilute 20.0 ml. of solution B to 100 ml. with ethanol.

(D) Dilute estriol solution, 0.2 μg./ml. Dilute 10.0 ml. of solution C to 100 ml. with ethanol.

These solutions are stable indefinitely if stored at 4°C.

Hydrolysis and Extraction

Dilute the 24-hour urine to 2500 ml. with distilled water and mix thoroughly. If the original volume is greater than 2500 ml. dilute to the nearest 100 ml. mark. Into each of two glass-stoppered test tubes (30–40 ml. capacity) measure 5.0 ml. volumes of the urine. Add 5.0 ml. of distilled water and 1.5 ml. of concentrated HCl to each, mix thoroughly, and place the loosely stoppered tubes in a boiling water bath for exactly 60 minutes to hydrolyze estrogen conjugates. Cool rapidly in running water and add 4.0 ml. of 10 N NaOH with thorough mixing. Transfer the alkaline solutions thus obtained as completely as possible to 60 ml. separatory funnels via small glass filter funnels with wide stems for convenience.

NOTE 5: Do not use grease on the separatory funnel stopcocks. Moisten both stoppers and stopcocks with distilled water prior to use.

Rinse each test tube with 10 ml. of a mixture of equal volumes of benzene and hexane. Transfer this via the filter funnels to the corresponding separatory funnels and extract the alkaline solutions by shaking approximately 100 times. Allow the phases to separate completely.

NOTE 6: When extraction of an aqueous solution with an organic solvent is referred to hereafter the above technique is implied. Shaking should not be too vigorous otherwise troublesome emulsions may be formed. In the particular extraction described above the benzene:hexane mixture removes neutral compounds which might interfere in estrogen measurement. A small amount of estrogen (mainly estrone) is also extracted under these conditions.

Run off the aqueous alkaline phases completely into their original tubes and extract the organic phases remaining with 2 × 6 ml. of 1 N NaOH. Combine these NaOH extracts with the alkaline solutions already in the tubes and reject the organic extracts.

NOTE 7: Extraction with 1 N NaOH allows recovery of the estrogen fraction initially removed by the benzene:hexane.

To each tube add 2.4 ml. of concentrated HCl (the solution should still be alkaline to test paper at this stage) and 1 g. of solid NaHCO₃ stirring until solution is complete. Check the pH with narrow range test paper and ascertain that it is in the range 7.5–8.5. If this has not been reached add more NaHCO₃.

NOTE 8: The naturally occurring phenolic estrogen metabolites have pK values such that at pH values below 9.3 these compounds may be quantitatively extracted from aqueous solutions with ether. This step brings about considerable purification of the estrogen fraction leaving much of the contamination in the aqueous phase.

Extract the aqueous solution with 1×20 ml. and 2×10 ml. of ether, reject the aqueous phase and combine the estrogen-containing ethereal extracts in their separatory funnels. Wash each once with 8 ml. of carbonate buffer pH 10, once with 6 ml. of 8% $(NH_4)_2SO_4$ solution and twice with 4 ml. of distilled water.

NOTE 9: Washing an organic extract with an aqueous solution implies shaking gently 10–20 times and allowing the phases to separate. Carbonate buffer removes further impurities without loss of estrogen since, although the pH is 10, the ionic strength of the solution is sufficient to prevent this. A wash with $(NH_4)_2SO_4$ solution preserves the ionic strength at this stage. An immediate water wash would result in a high pH and a low ionic concentration—conditions favoring estrogen loss from the organic phase.

Reject the aqueous washings and distil the ether to dryness from 100 ml. round-bottom flasks on a steam bath, adding a clean glass bead to prevent bumping. Remove residual ether vapor with a gentle stream of filtered air or nitrogen while the flasks are still warm.

Development of Fluorescence

Completely dissolve the residue in each flask in 3.00 ml. of ethanol. Remove 1.00 ml. or 2.00 ml. by pipette and transfer to glass-stoppered tubes (approximate capacity 10 ml.) containing 20 ± 5 mg. of powdered hydroquinone. The size of aliquot depends upon the approximate estrogen level expected. Into similar tubes containing hydroquinone measure duplicate 0.25 ml. (0.05 μg.) or 0.50 ml. (0.1 μg.) volumes of dilute estriol standard D when tetrabromoethane is to be used or twice these amounts in the case of tetrachloroethane, again depending on the estrogen level expected. A blank tube contains only hydroquinone. Evaporate the solutions to dryness in a water bath at approximately 60°C. under a gentle stream of filtered air or nitrogen. To each tube add 0.60 ml. of distilled water from a burette followed by 1.10 ml. of concentrated H_2SO_4 and mix while cooling in tap water. Place the loosely stoppered tubes in a *boiling* water bath for 40 minutes shaking twice during the first few minutes of heating. (At this point place the *p*-nitrophenol solution in ice-water in readiness for use). Cool the tubes in tap water and place in ice-water for at least 3 minutes. Add to each from a burette (carefully down the side of the tube so as to form a separate layer) 2.25 ml. of distilled water and cool at 0°C. for 3 minutes prior to mixing thoroughly then return to the ice-water for a further 3 minutes period. Add to each tube from a burette 3.00 ml. of *p*-nitrophenol solution, place in ice-water for 3 minutes then shake the stoppered tubes vigorously about 100 times. Centrifuge for 4 minutes at 3–4 × 10³ r.p.m., aspirate the upper (aqueous) phases, return the tubes containing the fluorescent extracts to the ice-water bath and protect from direct light in readiness for fluorimetry. Readings should be made as soon as possible thereafter.

NOTE 10: Protection from light at 0°C. improves the stability of the fluorescence so that no fading is observed within at least 1 hour.

Fluorimetry

Use a fluorimeter containing a photomultiplier, with an interference filter transmitting the Hg 546 mμ (green) line in the primary position and a glass filter (orange) transmitting maximally at about 585 mμ in the secondary position. These filters are particularly suitable when tetrabromoethane is the solvent since maximum absorption occurs at 543 mμ.

NOTE 11: In the authors' laboratory a Farrand Model A Fluorometer has been used mainly with an orange glass filter 4 mm. in thickness (Schott OG 2) in the secondary position. Several types of instrument are of course available, and the origin of the filters will depend on the laboratory concerned.

Set the galvonometer scale to 100% deflection with the standard eosin solution and read blanks, unknowns, and standards against this in matched glass tubes of capacity 2–3 ml. and of light path 1.00 cm.

NOTE 12: Using the Farrand fluorometer with light aperture 6 the reagent blank shows a deflection of 2–4 scale divisions under these conditions. Most normal estrogen values may be read from this scale. When tetrabromoethane is the solvent a linear relationship between fluorescence intensity and concentration of estriol exists over the range 0.01 μg. (approximately 10% scale deflection; including reagent blank) to 0.10 μg. (approximately 70% scale deflection). Similar readings are found for 0.02–0.20 μg. of estriol when tetrachloroethane is the solvent. For urines of very low estrogen content the scale may be set with a more dilute eosin solution. For high values the original eosin standard may be set to read 50% scale deflection.

Calculation

Micrograms of total estrogen (in terms of estriol) per 24 hours.

$$= \frac{F.R.u}{F.R.s} \times S \times \frac{3}{v} \times \frac{2500}{5}$$

where F.R.u = fluorescence intensity of the unknown corrected for the blank reading.

F.R.s = fluorescence intensity of S μg. of estriol standard corrected for the blank reading.

v = volume of ethanolic estrogen extract taken for fluorimetry.

Ib. TOTAL ESTROGENS IN PREGNANCY URINE

Reagents

Reagents 1–15 as described in Method Ia.
16. *Clinitest tablets for semiquantitative estimation of urinary sugar.*

Standard Estrogen Solutions

Dilute estriol solutions C and D as described in Method Ia.

Procedure

Test all urines by the Clinitest method and treat those which are positive (even a trace of sugar) by the following procedure. Also use the following analytical method for urines collected during the first 3–4 months of pregnancy or for any urines suspected of being very low in estrogen content.

Dilute the 24-hour urine to 2500 ml. as in Method Ia and measure duplicate 0.50 ml. (for sugar-containing urines of mid or late pregnancy) or 1.00 ml. (for all urines of early pregnancy) into glass-stoppered tubes of 30–40 ml. capacity. Dilute to 10.0 ml. with distilled water and mix thoroughly with 1.5 ml. of concentrated HCl.

NOTE 13: Before acid hydrolysis, 10- to 20-fold dilution of urine protects against the destructive influence of glucose on urinary estrogens (9, 10). Even in the absence of sugar urine concentration is an important factor in estrogen destruction during acid hydrolysis (10). In pregnancy urines these dilutions can be performed due to high estrogen titers.

Proceed exactly as described in Method Ia from the hydrolysis stage to the preparation of the dried ether extracts. Dissolve the extracts in 3.00 ml. of ethanol and proceed as described for the development of fluorescence in Method Ia.

In early pregnancy (up to 3–4 months) take 0.50–2.00 ml. of the ethanolic extracts and measure fluorimetrically using as standard duplicate 0.05 or 0.10 μg. of pure estriol as in Method Ia.

In urines of later pregnancy take similar aliquots and measure the estrogen content colorimetrically as described below. As standard use duplicate 1.00 ml. volumes (2 μg.) of dilute estriol solution C (Method Ia).

In sugar-free urines of mid or late pregnancy measure total estrogens directly on the urine. For this purpose measure duplicate 0.150 ml. volumes of urine into glass-stoppered tubes of 10 ml. capacity each containing 20 ± 5 mg. of hydroquinone. Dilute with 0.45 ml. of distilled water and proceed exactly as described for the development of fluorescence in Method Ia commencing with the addition of 1.10 ml. of concentrated H_2SO_4.

NOTE 14: It has been found in the authors' laboratory that the presence of glucose in urine produces a pink color with a broad spectrum (λ max. = 510 mμ) in the final extract which masks any color given by the estrogens (λ max. = 538 mμ). Glucose also exerts a destructive effect on estrogens in hot concentrated sulfuric acid solutions. Thus sugar-containing urines should be analyzed by the rather longer method involving preliminary splitting of estrogen conjugates in highly diluted urine.

Colorimetry

Transfer the tetrabromo- or tetrachloroethane extracts to 3 ml. glass cells of 1.00 cm. light path and read unknowns and standards against a reagent blank prepared as in Method Ia. Measure the absorbance at 510, 543, and 576 mμ (tetrabromoethane) or 506, 538, and 570 mμ (tetrachloroethane) in a spectrophotometer reading in the visible range. There is linearity between absorbance and estriol concentration up to 10 μg. but under the analytical conditions it is unlikely that more than the lower half of this range will be used in practice.

Calculations

Calculate the corrected absorbance (A corr.) for unknowns and standards as follows:

$$(\text{tetrabromoethane}) \text{ A corr.} = A_{543} - \frac{A_{510} + A_{576}}{2}$$

$$(11)$$

$$(\text{tetrachloroethane}) \text{ A corr.} = A_{538} - \frac{A_{506} + A_{570}}{2}$$

Note 15: Correction of the absorbance at the peak wavelength (538 or 543 mμ) corrects for interference by nonspecific chromogens provided that the latter exhibit a linear spectrum over the wavelength range measured (in this case 506–570 or 510–543 mμ) (11), which is generally so in the absence of interfering medication, etc. (See Note 1).

Note 16: Tetrachloroethane is much the preferred solvent because it causes no turbidity whereas the tetrabromoethane does. The extinctions are the same, approximately.

Milligrams of total estrogen (in terms of estriol) per 24 hours when the longer method involving hydrolysis and extraction is employed

$$= \frac{\text{A corr.(u)}}{\text{A corr.(s)}} \times 2 \times \frac{3}{v} \times \frac{2500}{V} \times \frac{1}{1000}$$

where A corr.(u) = corrected absorbance at 538 or 543 mμ of the unknown

A corr.(s) = corrected absorbance at 538 or 543 mμ of the standard (2 μg. of estriol)

v = volume of ethanolic estrogen extract taken for colorimetry

V = volume of urine originally taken for analysis (0.5 or 1.0 ml.)

Milligrams of total estrogen (in terms of estriol) per 24 hours using the direct method without prior purification

$$= \frac{\text{A corr.(u)}}{\text{A corr.(s)}} \times 2 \times \frac{2500}{0.15} \times \frac{1}{1000}$$

where the symbols have the same meaning as above.

II. ESTRIOL IN PREGNANCY URINE

Reagents

Reagents 1–3, 6–9, and 11–13 as in Method Ia.

17. *Sodium hydroxide, 2 N.* Dilute 200 ml. of 10 N NaOH (reagent

2, Method Ia) to 1 l. with distilled water. Store in a polyethylene bottle.

18. Kober color reagent (7). Mix carefully (with stirring and cooling) 760 ml. of reagent grade concentrated H_2SO_4 with 200 ml. of distilled water. Cool to room temperature and dilute to 1 l. Dissolve 10 mg. of reagent grade sodium nitrate ($NaNO_3$) and 20 mg. of resublimed *p*-quinone in 1 l. of the diluted acid and warm in a water bath at 100°C. until a light green color just appears. Immediately add 20 g. of hydroquinone and heat in a boiling water bath for 45 minutes with occasional shaking until solution is complete. Allow to stand in the dark at room temperature for about 1 week then filter through a sintered glass (porosity 4) funnel. Store in a dark glass bottle at room temperature.

NOTE 17: The presence of sodium nitrate and *p*-quinone eliminates the variability in the reagent caused by using different brands or batches of sulfuric acid. The reagent as prepared should be straw-colored and is stable under the above conditions for several months. On occasion a slight precipitation may occur which may be removed by further filtration without detriment to the reagent.

Standard Estrogen Solutions

Dilute estriol solutions B and C as described in Method Ia.

Hydrolysis and Extraction

Dilute the 24-hour urine to 2500 ml. with distilled water and measure duplicate 10.0 ml. volumes into 250 ml. round-bottom flasks fitted with standard taper glass sockets. Dilute each to 100 ml. with distilled water, add a glass bead to prevent bumping and bring to boiling point under reflux condensers. As soon as boiling begins add 15 ml. of concentrated HCl down each condenser and boil for exactly 60 minutes. Cool the flasks in running water and transfer the contents completely to 250 ml. separatory funnels (see Note 5). Extract the urine with 1 × 30 ml. and 3 × 25 ml. of ether (see Note 6) using the first 30 ml. volume to rinse out the hydrolysis flask. Collect the ether extracts in 250 ml. Erlenmeyer flasks and return the combined extracts to their respective separatory funnels. Wash once with 20 ml. of concentrated carbonate buffer (see Note 9) which has previously been used to rinse the respective Erlenmeyers and discard the aqueous phase. Add 5.0 ml. of 2 N NaOH to each separatory funnel and shake 100 times then, *without rejecting the aqueous phase*, add 20 ml. of 8% $NaHCO_3$ and repeat the shaking. Test the pH of the aqueous phase with test paper and if no higher than 10 discard it. If the pH is too high add more $NaHCO_3$ and shake again until pH 10 is reached.

NOTE 18: Shaking with 2 N NaOH causes formation of pigments and results in some extraction of estrogen into the aqueous phase. Lowering of the pH to 10 leaves the pigmented impurities in the aqueous phase but causes partition of the estrogen into the ether due to considerations of ionic strength.

Wash the ether extracts with 1 × 5 ml. of 8% $NaHCO_3$ in order to lower pH without decreasing ionic strength (see Note 9) and discard the aqueous phase. Finally wash with 2 × 5 ml. of distilled water using this first of all to rinse the stoppers and inlet walls of the separatory funnels and the small glass filter funnels used for transfer. Discard the aqueous washings, run the ether extracts into 250 ml. round-bottom flasks fitted with standard taper sockets and distil to dryness on a steam bath using glass beads to prevent bumping. Remove the last traces of ether vapor from the warm flasks with a gentle current of filtered air or nitrogen.

Dissolve the estrogen-containing residues from the ether distillations in 1.5 ml. of ethanol and transfer completely to 125 ml. separatory funnels with 10, 10, and 5 ml. portions of benzene.

NOTE 19: Estriol is relatively insoluble in benzene so that it is necessary to dissolve the residue in ethanol prior to diluting with benzene.

Rinse the round-bottom flasks with 25 ml. of distilled water and add this to the respective separatory funnels. Shake the funnels some 60 times to ensure equilibration without producing emulsions. Collect the aqueous phase, after separation of the liquid layers, in a 250 ml. round-bottom flask. Wash the benzene further with 1 × 25 ml. and 2 × 12.5 ml. of water and add these to the flask. Discard the benzene.

NOTE 20: The partition coefficient of estriol in the system benzene/water is such that this steroid may be quantitatively recovered in the aqueous phase leaving most of the other known estrogen metabolites in the benzene. A certain fraction of the naturally occurring metabolite 16-epiestriol will remain in the aqueous phase but because of the very high ratio of estriol:16-epiestriol in pregnancy urine this is of little practical importance.

Add to each estriol-containing aqueous solution 7.5 ml. of 10 N NaOH and allow to stand at room temperature for 20 minutes.

NOTE 21: In the original Bauld method (4) as applied to nonpregnancy urine, hot saponification is carried out at this stage. However, in pregnancy urine where the ratio estriol:impurities is high this is unnecessary.

Transfer each aqueous extract to a 250 ml. separatory funnel and extract with 1 × 100 ml. of ether. Discard the ether layer.

NOTE 22: Under these conditions extraction with ether removes neutral hydrophilic compounds but not estriol.

Adjust the alkaline solutions, in their separatory funnels, to pH 9 by passing CO_2 gas by way of a manifold through 1 mm. capillary tubes dipping into the solutions (see Note 8). Check the pH periodically with narrow range test paper until the desired value is reached. Extract the aqueous solutions with 4 × 40 ml. of ether using the first volume to wash the capillary tubes. Collect the estriol-containing ether extracts in 250 ml. Erlenmeyer flasks and when the extraction has been completed combine these extracts in their respective separatory funnels. Wash each with 1 × 5 ml. of 8% $NaHCO_3$ and 2 × 5 ml. of distilled water discarding the aqueous washings. Use the water washings first of all to rinse separatory funnel stoppers, inlet walls, etc., as described above. Distil the ether extracts to dryness from 500 ml. round-bottom flasks on a steam bath using glass beads to prevent bumping. Remove residual ether vapor from each flask while still warm with a gentle current of filtered air or nitrogen.

Color Development

Completely dissolve the dry residue in each flask in 3.00 ml. of ethanol and remove suitable aliquots from each for analysis. During the first 3 months of pregnancy remove 1.00–2.00 ml. volumes and analyze by the colorimetric method described under Method Ib using 2 μg. of estriol as standard. In later pregnancy remove 2.00 ml. (fourth month) to 0.10 ml. (eighth to ninth month) and evaporate in test tubes of uniform dimensions (length 15 cm. diameter 2.5 cm.) in the presence of 50 ± 5 mg. of hydroquinone. As standards use duplicate 0.50 ml. volumes (5 μg. of estriol) of dilute estriol solution B (Method Ia) similarly evaporated and as blank a tube containing only hydroquinone.

NOTE 23: Hydroquinone is required at this stage to protect the estrogen against the destruction which occurs during heating in the presence of H_2SO_4 and urinary residues.

Add 2.60 ml. of Kober color reagent to each tube from a burette and heat for 20 minutes in a *boiling* water bath shaking the tubes briefly during the third and tenth minutes to ensure solution of the hydroquinone. Cool to room temperature in a cold water bath and add to each tube 50 ± 5 mg. of hydroquinone.

NOTE 24: After 20 minutes of heating the estriol-containing solutions are a yellowish color with a green fluorescence. Additional hydroquinone is necessary to replace that sulfonated during heating. The solutions may safely be kept at this stage for 1–2 hours if necessary.

To each tube add 0.70 ml. of distilled water from a burette and mix by shaking. Heat again in a *boiling* water bath for 15 minutes shaking each tube during the third and tenth minutes. Cool to room temperature in cold water.

NOTE 25: Dilution with water followed by heating converts the fluorescent material to a stable form with a pink color (λ max. = 512.5 mμ). Colorimetry may be performed within the next 2 hours.

Colorimetry

Where low estriol levels are encountered and the tetrabromo- or tetrachloroethane extraction procedure is employed, perform colorimetry exactly as described in Method Ib reading at 3 wavelengths. Otherwise transfer the solutions to 3 ml. glass cells of light path 1.00 cm. and read unknowns and standards against a reagent blank set to zero absorbance at 480, 512.5, and 545 mμ in a visible light spectrophotometer. Linearity exists between absorbance and estriol concentration up to 10 μg.

Calculation

When the tetrabromo- or tetrachloroethane extraction method is used the calculation is done as described in Method Ib and milligrams of estriol per 24 hours.

$$= \frac{A \ corr.(u)}{A \ corr.(s)} \times 2 \times \frac{3}{v} \times \frac{2500}{10} \times \frac{1}{1000}$$

(see Method Ib for explanation of symbols)

When the older Kober method is performed in urines of higher estriol content the corrected absorbance for unknowns and standards is calculated as follows:

$$A \ corr. = A_{512.5} - \frac{A_{480} + A_{545}}{2} \ (\text{see Note 15})$$

Milligrams of estriol per 24 hours.

$$= \frac{A \ \text{corr.}(u)}{A \ \text{corr.}(s)} \times 5 \times \frac{3}{v} \times \frac{2500}{10} \times \frac{1}{1000}$$

where A corr.(u) = corrected absorbance at 512.5 mμ of the unknown

A corr.(s) = corrected absorbance at 512.5 mμ of the standard (5 μg. of estriol)

v = volume of ethanolic estrogen extract taken for colorimetry.

Possible Modification of the Methods

These methods may, and presumably will be modified in a number of ways in individual laboratories. A greater degree of sensitivity may be achieved by using smaller volumes of reagents for colorimetry and fluorimetry and by employing microcells for measurement. However, it does not necessarily follow that these modifications can always be made since the effect of solvent and urinary residues on smaller reagent volumes might result in high "background" interference. Additional means of increasing sensitivity include analyzing the entire estrogen-containing ethereal extract after direct evaporation in the tube used for development of color or fluorescence (2) rather than taking an aliquot as described above. It is also possible to omit dilution of the original urine volume to 2500 ml. although it is advisable to aim at a reasonably constant starting volume in order to minimize the variability in estrogen recovery caused by differences in urine concentration during hot acid hydrolysis (see Note 13). This is not so important for pregnancy urines where, because of the high estrogen levels, 10- or 20-fold dilution can be made without resulting in unwieldy working volumes. Sugar-containing urines from nonpregnant subjects might be best hydrolyzed by β-glucuronidase-containing enzyme preparations (9, 10, 12).

Accuracy of the Methods

The following data were obtained in the authors' laboratory by adding known amounts of pure estrogen to the requisite volume of nonpregnancy urine and calculating the recovery after applying the method in question. Recovery of 0.06–0.30 μg. of estriol added to 5 ml. of urine and treated by Method Ia = 89 ± 5.6% (S.D.); for

0.27 μg. of estrone the average recovery = 81%. Recovery of 1–4
μg. of estriol added to 0.5 or 1.0 ml. of urine diluted to 10 ml. with
water and treated by Method Ib (hydrolysis plus extraction followed
by the colorimetric technique) = 92 ± 4.9% or by Method Ib (direct
method) = 95 ± 5.9%. Recovery of 1–50 μg. of estriol added to 10
ml. of urine diluted to 100 ml. with water and treated by Method II
(either colorimetric method used) = 82 ± 5.0%.

Normal Values

It is extremely difficult to quote normal values for urinary estrogens
due to the considerable fluctuations which occur both during the
menstrual cycle and in pregnancy. In the nonpregnant state the
total estrogen level extends over the approximate range 10–100 μg./24
hours the lower values being characteristic of the postmenopausal
female. The highest levels are observed at about the fourteenth day
(ovulation peak) and the twenty-second day (luteal maximum) of
the menstrual cycle, while during the proliferative phase about one-
quarter to one-third of these values are found. There are considerable
differences in the amounts of estrogen metabolites excreted from one
individual to another. In the normal human male the levels are rather
similar to those encountered in the proliferative phase of the normal
menstrual cycle. During pregnancy the estrogen levels found in the
luteal phase of the normal menstrual cycle increase about one
thousandfold giving total values at term of about 40 mg./24 hours
on the average. Again there is a very considerable variation from one
individual to another. The excretion of estriol during pregnancy
parallels total estrogen very closely and accounts for 60–70% of the
total during the later stages. On delivery urinary estrogen levels fall
very rapidly.

Discussion

In the nonpregnant state the estrogen pattern in human urine is
essentially the same as that found when estrone or estradiol-17β is
administered to a human subject (13). It is generally accepted therefore that the estrogen secreted is either one or both of these compounds
(13, 14) and that the urinary metabolites arise as catabolic products
in the liver and probably other tissues. In the menstruating female
the ovary is the main source of estrogen while in the postmenopausal
woman and in the male the adrenal cortex may contribute substantially

to the urinary estrogen level (15, 16). In pregnancy the placenta is the main source of estrogen and probably produces estriol besides estrone and/or estradiol-17β (13, 17, 18).

In general, more meaningful results are obtained by measuring specific estrogen metabolites than by analyzing these as a group just as in the case of the neutral 17-ketosteroids. Thus in the total or group method a number of compounds of differing extinction coefficients or fluorescence intensities are measured in terms of a single estrogen (e.g., estriol). Frequently, however, *strict* quantification of estrogen levels is less important than the trend in excretion over a certain period of time. Therefore, in view of the intricacy of separating steroid metabolites there is much to be said in favor of the total estrogen methods in routine clinical chemistry laboratories. This is particularly true of the nonpregnant state where no single metabolite represents a nearly quantitative picture. It has already been shown by Ittrich (2) that during the menstrual cycle the pattern of total estrogen excretion is qualitatively the same as that reported by Brown (19) for estrone, estradiol-17β, and estriol and therefore presumably reflects ovarian function. Also, the sum of the latter 3 compounds when measured separately by a reliable method accounts quite consistently for 70–80% of the total values (2).

In the pregnant state estriol is quantitatively by far the most important urinary estrogen and the measurement of this alone may be a useful test of placental efficiency (20). The relationship between urinary estriol and total estrogens during the last 20 weeks of pregnancy is reasonably consistent. Thus estriol expressed as a percentage of total estrogens measured by the direct Ittrich method is 59 ± 8.7% (S.D.) and when the Ittrich method including hot acid hydrolysis and extraction is employed is 67 ± 12.6% (21). The direct Ittrich method is more accurate since it does not result in a large loss of the ketolic estrogen fraction (16α-hydroxyestrone, 16-ketoestradiol-17β, and 16β-hydroxyestrone) which is quantitatively important in human pregnancy urine (22, 23, 24). On the other hand the Ittrich method involving preliminary hot acid hydrolysis results in large losses (40–60%) of these ketolic compounds (24) so that no strict comparison can be made of results obtained by the two Ittrich procedures.

The methods described above are sufficiently sensitive and precise for use in the clinical chemistry laboratory. Moreover they are capable of being performed within a working day, several duplicate analyses being possible. The reliability of each method must be thoroughly

82   R. HOBKIRK AND A. METCALFE-GIBSON

tested in the individual laboratory prior to being employed for routine analysis. It should be borne in mind that a single estrogen value is seldom of much significance and that serial analyses are always necessary to obtain worthwhile data.

REFERENCES

1. Ittrich, G., Eine neue Methode zur chemischen Bestimmung der östrogenen Hormone im Harn. Z. Physiol. Chem. 312, 1–14 (1958).
2. Ittrich, G., Untersuchungen uber die Extraktion des roten Kober-Farbstoffes durch organische Losungsmittel zur Östrogenbestimmung im Harn. Acta Endocrinol. 35, 34–48 (1960).
3. Kober, S., Eine kolorimetrische Bestimmung des Brunsthormons (Menformon). Biochem. Z. 239, 209–212 (1931).
4. Bauld, W. S., A method for the determination of oestriol, oestrone and oestradiol-17β in human urine by partition chromatography and colorimetric estimation. Biochem. J. 63, 488–495 (1956).
5. Bauld, W. S., and Greenway, R. M., Chemical determination of estrogens in human urine. In "Methods of Biochemical Analysis" (D. Glick, ed.) Vol. 5, pp. 337–406. Wiley (Interscience), New York, 1957.
6. Hobkirk, R., Measurement of urinary estrogens. Clin. Chem. 7, 469, 1961.
7. Bauld, W. S., Some errors in the colorimetric estimation of oestriol, oestrone and oestradiol by the Kober reaction. Biochem. J. 56, 426–434 (1954).
8. Brown, J. B., Bulbrook, R. D., and Greenwood, F. C., An additional purification step for a method for estimating oestriol, oestrone and oestradiol-17β in human urine. J. Endocrinol. 16, 49–56 (1957).
9. Hobkirk, R., Alfheim, A., and Bugge, S., Hydrolysis of estrogen conjugates in diabetic pregnancy urines. J. Clin. Endocrinol. Metabolism 19, 1352–1356 (1959).
10. Brown, J. B., and Blair, H. A. F., The hydrolysis of conjugated oestrone, oestradiol-17β and oestriol in human urine. J. Endocrinol. 17, 411–424 (1958).
11. Allen, W. M., A simple method for analyzing complicated absorption curves, of use in the colorimetric determination of urinary steroids. J. Clin. Endocrinol. 10, 71–83 (1950).
12. Hobkirk, R., Bugge, S., and Nilsen, M., Hydrolysis of estrogen conjugates in human urine. Proc. 4th Inter. Congr. Clin. Chem. (1960). p. 164.
13. Brown, J. B., The relationship between urinary estrogens and estrogens produced in the body. J. Endocrinol. 16, 202–212 (1957).
14. Fishman, J., Bradlow, H. L., and Gallagher, T. F., Oxidative metabolism of estradiol. J. Biol. Chem. 235, 3104–3107 (1960).
15. Brown, J. B., Falconer, C. W. A., and Strong, J. A., Urinary estrogens of adrenal origin in women with breast cancer. J. Endocrinol. 19, 52–63 (1959).
16. Givner, M. L., Bauld, W. S., Hale, T. R., Vagi, K., and Nilsen, M., The effect of corticotropin, chorionic gonadotropin and testosterone propionate on urinary estradiol-17β, estrone and estriol of human subjects with previous myocardial infarction. J. Clin. Endocrinol. Metabolism 20, 665–674 (1960).
17. Pincus, G., and Pearlman, W. H., The intermediate metabolism of the sex hormones. Vitamins Hormones 1, 293–343 (1943).

18. Hobkirk, R., Blahey, P. R., Alfheim, A., Raeside, J. I., and Joron, G. E., Urinary estrogen excretion in normal and diabetic pregnancy. *J. Clin. Endocrinol. Metabolism* **20**, 805–813 (1960).
19. Brown, J. B., Urinary excretion of oestrogens during the menstrual cycle. *Lancet* **I**, 320–323 (1955).
20. Zondek, B., and Goldberg, S., Placental function and foetal death. II. Urinary oestriol excretion test in advanced pregnancy. *J. Obst. Gynec. Brit. Emp.* **64**, 1–13 (1957).
21. Hobkirk, R., unpublished work.
22. Layne, D. S., and Marrian, G. F., The isolation of 16β-hydroxyestrone and 16-oxo-estradiol-17β from the urine of pregnant women. *Biochem. J.* **70**, 244–248 (1958).
23. Hobkirk, R., and Nilsen, M., Observations on the occurrence of six estrogen fractions in human pregnancy urine. I. Normal pregnancy. *J. Clin. Endocrinol. Metab.* **22**, 134–141 (1962).
24. Hobkirk, R., and Nilsen, M., Observations on the occurrence of six estrogen fractions in human pregnancy urine. II. Diabetic pregnancy. *J. Clin. Endocrinol. Metab.* **22**, 142–146 (1962).

ADDITIONAL REFERENCE

Brown, J. B., The determination and significance of the natural estrogens. *In* "Advances in Clinical Chemistry" (Sobotka and Stewart, eds.) Vol. 3, pp. 157–233. Academic Press, New York, 1960.

MEASUREMENT OF TOTAL ESTERIFIED FATTY ACID AND TRIGLYCERIDE CONCENTRATIONS IN SERUM*

Submitted by: John G. Reinhold, Virginia L. Yonan, and Edythe R. Gershman, William Pepper Laboratory of Clinical Medicine, Hospital of the University of Pennsylvania, Philadelphia, Pennsylvania

Checked by: Nelson Young, Department of Biochemistry, Medical College of Virginia, Richmond, Virginia and
Ruth McNair, Providence Hospital, Detroit, Michigan

Introduction

The measurement of total esterified fatty acid (TEFA) and of triglyceride concentrations in body fluids has presented many difficulties, and because of this, neither measurement has attained the extensive application that its importance warrants. Instead, physicians usually have been obliged to depend upon serum cholesterol, and to a lesser extent, phospholipid measurements for information concerning disturbances of lipid metabolism. It is well established that triglyceride concentrations may change quite independently of the other major lipid components in plasma (1). As a result, significant changes may be overlooked unless triglyceride measurements are included in lipid studies.

Until recently, only gravimetric (2), titrimetric (2, 3, 4), and oxidative (5, 6) methods have been available for measurement of total fatty acid and, indirectly, of triglyceride concentrations. The volume of serum required for the gravimetric and titrimetric methods, 3–5 ml., is relatively large. Each of these methods have objectionable features. Saponification, neutralization, filtration, and washing, as well as the final quantification by weighing, titration, or oxidation, require special care and considerable time.

The method described in the present paper, based on conversion of esterified fatty acids to ferric hydroxamates, offers many advantages over the older procedures. It has been used by the authors during the past 7 years with satisfactory results for many thousands of analyses.

The ferric hydroxamate method owes its development to an observa-

* Based on the method of Stern and Shapiro (10) and Nailor et al. (11).

tion by Feigl et al. (7) that esters and anhydrides reacted with hydroxylamine to form hydroxamic acids. The latter, when treated with ferric ions, yielded intense red colors suitable for "spot" reactions. Hill (8) adapted these reactions to the measurement of free and esterified fatty acids in metal coatings. The free acids were first converted to methyl esters. Bauer and Hirsch (9), devised a procedure for total esterified fatty acids based on the reaction with hydroxylamine. This was simplified by Stern and Shapiro (10), who found that the coupling with hydroxylamine would occur in the alcohol-ether solutions used to extract lipids from serum. Others carried out the reaction in ether under anhydrous conditions. Removal of water added appreciably to the time required for analysis. The color formed per mole of ester is increased when anhydrous conditions are used, yet under the conditions described by Stern and Shapiro (10) the yield of color is sufficient to enable 0.1-ml. samples of serum to be used. However, the reaction can be carried out in the presence of water only if the concentration of alcohol provided is sufficient to keep the long chain hydroxamic acids in solution.

The method to be presented is based mainly on that of Stern and Shapiro (10). It incorporates certain modifications from Nailor et al. (11). Other ferric hydroxamate methods have been described by Kornberg and Pricer (12), Jarrier and Polonovski (13), Jonnard (14), Gey and Schön (15), and Rosenthal et al. (16).

The reactions occurring are outlined by Feigl (7), Hill (8), and by Goddu et al. (17), as:

$$RCOOR + NH_2OH \xrightarrow{(OH)^-} RC:ONHOH + ROH$$

$$1/n\ FE^{3+} + RC:ONHOH \longrightarrow \begin{array}{c} RCNH+H^+ \\ \parallel\ | \\ O\ O \\ Fe/n \end{array}$$

Bayer and Reuther (18), find that three different colored complexes may form depending on molar ratios of the reactants, pH and solvent. The reaction is not confined to derivatives of fatty acids. Similar reactions occur with acid chlorides and anhydrides. Hestrin (19) described the reaction of hydroxylamine and acetylcholine, and Jandorf (20), a different type of reaction with nerve gases. Esters of amino acids react in a similar way (21). The amounts of reactive substances other than fatty-acid esters extracted by alcohol-ether from blood serum are, so far as is known, too small to cause significant elevation of fatty acid concentrations. The effects of various substituents on the reaction have been studied by Goldenberg and Spoerri (22). Conditions under which hydroxamic acids may form with amides and other carboxylic acid derivatives are described by Wainfan and van Bruggen (23).

Reagents

1. *Alcohol-ether (Bloor's reagent).* Three volumes of ethanol (99.5% or better) are mixed with 1 volume of ethyl ether. The ethanol should give a low blank. Redistillation may be necessary; however it has been the authors' experience that industrial absolute alcohol has not required further purification. Gey and Schön (15) finds that aldehyde and peroxide impurities may be major sources of error.

NOTE: Ethyl ether of analytical reagent quality has been used by the authors with good results, precautions being used to forestall peroxide formation. Some brands of ether for anesthesia give extremely high blanks, and ether in tins should be avoided (24).

2. *Hydroxylamine hydrochloride, 2M.* Dissolve 14 g. of hydroxylamine hydrochloride (reagent quality) in water and dilute to 100 ml. The solution is kept in the refrigerator where it remains stable for at least 1 month. It should be warmed to room temperature before using.

NOTE: Hydroxylamine hydrochloride should be stored in tightly stoppered containers. Aged or discolored material should be rejected.

3. *Sodium hydroxide solution, 3.5 N.* Standardized by titration. The presence of Na_2CO_3 contamination is not objectionable [Hill (8)].

4. *Hydrochloric acid, 4.0 N.* One volume of concentrated acid (sp. gr., 1.18) is added to 2 volumes of water and standardized by titration.

5. *Ferric perchlorate, 0.37 M.* Dissolve 13 g. of ferric perchlorate (11) (G. Frederick Smith of Chemical Co., Columbus, Ohio) in 0.1 N HCl and dilute to 100 ml. The solution should be kept in a brown bottle. The 0.1 N HCl may be prepared with sufficient exactness by diluting 0.8 ml. of concentrated HCl to 100 ml.

6. *Standard solution of fatty acid esters.* Triolein is preferred as a standard for study of human body fluids, although other higher fatty acid esters may be substituted. Pure triolein may be obtained from the Hormal Foundation, Austin, Minn. To prepare a standard solution containing 40 meq./l., dissolve 0.590 g. in alcohol-ether and adjust the total volume to 50 ml. Triolein gradually deteriorates and should be replaced yearly. Corn or olive oil may be used in place of triolein.

7. *Working standards* are prepared by diluting 5 ml. of the standard solution described above to 200 ml. in a volumetric flask using alcohol-ether. (One ml. contains 1 μeq. of esterified fatty acid.)

8. *Alternative standard, cholesteryl acetate.* Dissolve 0.215 g. of cholesteryl acetate [Rosenthal *et al.* (16)] in alcohol-ether and dilute to 200 ml. in a volumetric flask. (One ml. contains 2.5 μeq. of esterified acid.

A working standard is prepared by diluting 20 ml. of stock standard to 100 ml. with alcohol ether. (One milliliter contains 0.5 μeq.)

Procedure

Serum should be separated promptly from blood collected at least 12 hours after the last meal. An overnight fasting period is convenient. The serum should be added to the alcohol-ether without delay because of the lipolytic activity that occurs in some specimens.

Alcohol-ether extracts of serum may be stored for several days without change. Analyses are done in duplicate. Not more than twelve to twenty samples are treated with ferric perchlorate at a time, the number being dependent upon the skill and speed of the analyst.

Measure 0.1 ml. of serum[1] into a glass-stoppered test tube or cylinder graduated at 10 ml. Approximately 8 ml. of alcohol-ether are added along the wall of the container so as to disturb the serum as little as possible. The container is then tapped sharply a number of times to produce a fine dispersion of the serum. It is placed in a water bath with the temperature regulated at approximately 65°C. for 1 hour.[2] The contents of the cylinder are cooled to 25°C. and sufficient alcohol-ether is added to bring the volume to 10.0 ml. After mixing, the solution is filtered through paper (Whatman No. 1 has been used). Rapid filtration is desirable. The funnel is covered with a watch glass to prevent evaporation.

---

[1] Instead of 0.1 ml., 0.2 or 0.3 ml. of serum may be used when specimens with low concentrations of fat are examined. However, the authors have found the use of 0.1-ml. samples of human serum to be preferable as a routine, since turbidity is less frequently encountered than when larger volumes are used.

[2] Stern and Shapiro merely heated the serum alcohol-ether mixture to the boiling point but the procedure described in the present paper resulted in higher yields of TEFA. Constant temperature baths maintained at 56°C are available in most laboratories and these may be used.

Moistening the ground-glass stopper with water will prevent leakage of alcohol-ether when mixing [Sperry (47)].

Transfer 3.0 ml. of the filtrate to a large test tube of 20 to 25 mm. diameter.[3] If the minimal volume required for the photocolorimeter exceeds 5 ml., sufficient alcohol-ether should be added at this stage to produce the desired total volume.

Standards are prepared by measuring 0.2, 0.5, and 1.0 ml. of triolein working standard into similar tubes and adding sufficient alcohol-ether to make the volume equal to that of the serum filtrate. These are equivalent to 0.2, 0.5, and 1.0 μmole, respectively.

An additional tube containing alcohol-ether only is carried through the procedure to provide a blank.

To filtrates, standards, and blank are added 0.5 ml. of hydroxylamine hydrochloride solution. This and the following reagents may be added from burets. After mixing, 0.5 ml. of 3.5 N NaOH solution are added and the solutions are mixed again. The tubes are stoppered and allowed to stand at room temperature for 20 minutes to enable the hydroxylamine to react.[4] Add 0.6 ml. of 4 N HCl and mix the solution.[5] This is followed by the addition of 0.5 ml. of ferric perchlorate. Again the solutions are thoroughly mixed. The tubes are placed in a bath at 25°C. Five minutes should be allowed between time of mixing and measurement of absorbances. Exact timing is important.

Absorbances are measured at 515–530 mμ. Fading of the color, if it occurs, is compensated for by readjusting the zero setting before each measurement.

Calculation

Using the standard equivalent to 0.5 μeq. TEFA,

$$\text{meq./l} = A_u \times \frac{0.5}{A_s} \times \frac{10}{3} \times \frac{1000}{0.1} \times \frac{1}{1000}$$

$$= A_u \times \frac{5}{0.3A_s}$$

[3] The writers use cuvettes designed for the Evelyn photocolorimeter. These measure 22 × 250 mm.

[4] The time required for maximal coupling varies with different esters and with the conditions established in the test. The adequacy of the 20-minute period should be verified from time to time.

[5] The acidity after neutralization of NaOH is important. In the method as described it slightly exceeds 0.1 M. In this range color is more intense and more stable.

where A_u and A_s are the absorbances of serum filtrate and standard, respectively. If the higher or lower standards are used, appropriate adjustments are made.

ESTIMATION OF TRIGLYCERIDE CONCENTRATION

The difference between total esterified fatty acid (TEFA) and phospholipid (PFA) and fatty acids esterified in the cholesterol (CFA):

$$\text{Triglyceride fatty acid meq./l.} = \text{TEFA} - (\text{PFA} + \text{CFA})$$
$$\text{PFA} = \frac{\text{Lipid P} \times 10 \times 1.8}{31} = 0.58 \text{ Lipid P}$$

where lipid P is expressed as mg. of P per 100 ml., and 1.8 is the average molar ratio of fatty acid to phosphorus.

$$\text{CFA} = 0.29 \frac{\text{EC} \times 10}{387} = 0.0075 \text{ EC or } 0.0075 \times 0.70 \text{ TC} = 0.00525 \text{ TC}$$

where the esterified cholesterol (EC) or total cholesterol (TC) is expressed in mg. per 100 ml.;[6] 0.29 is an experimentally derived factor correcting for the incomplete reaction of higher fatty acids esterified with the cholesterol under the conditions prescribed.

Comments on the Method

CALCULATION OF TRIGLYCERIDE

An attempt was made to eliminate the need for phospholipid P measurements for the purpose of calculating triglyceride concentrations by treatment of alcohol-ether extracts of serum with Doucil, the adsorbent used by Van Handel and Zilversmit (25) for adsorption of phospholipid. This was not consistently successful and was abandoned.

The correction applied to the TEFA for phospholipid based on lipid P measurements must take into account the occurrence of sphingomyelin, a phospholipid containing only 1 eq. of fatty acid per mole of P. Man and Peters used the factor 0.58 which we have adopted. This assumes that sphingomyelin P represents 10% of the total. Some recent work [Phillips (26)] suggests that this figure may be too low, but its use seemed permissible pending final definition of serum phospholipid composition. Rapport and Alonzo (24) found that fatty acids of phospholipids reacted according to theory.

---

[6] Because of the constancy of the ratio of esterified to total cholesterol, it is our practice to calculate the esterified cholesterol correction on the assumption that it is 0.70 of the total cholesterol, except in patients with severe liver disease.

Kaplan (27) found and Entenman (28) confirmed that cholesterol esters of higher fatty acids did not yield hydroxamate derivatives when tested by the Stern and Shapiro method. We have found that cholesteryl palmitate, stearate, and oleate yield some color, but that this could not be measured accurately in our experiments because turbidity was always present. Clear, colored solutions were obtained when cholesteryl laurate and benzoate were used. A correction to be applied to the total esterified fatty acids for fatty acid combined as esters of cholesterol was based on experiments with the latter. The color yield from these esters was 0.29 that of an equivalent concentration of fatty acid combined as triglyceride. Since this factor is derived from esters different from those that predominate in blood serum, it should be regarded as tentative. Goldenberg and Spoerri (22) concluded that bulky substituents inhibited hydroxamic acid formation and it may be such an effect of cholesterol and the larger fatty acid molecules that impeded the reaction with hydroxylamine.

Various triglycerides also are said to yield different molar extinction values [Goddu et al. (17)] although others found no differences (29). In our work, tributyrin was used initially as a standard. However, the results were more reproducible when methyl oleate was substituted, and still better with triolein. Van Handel and Zilversmit (25) have used corn oil as a standard in their method. The use of natural food oils as standards appears to be permissible and to provide a simple answer to the search for a standard.

Color instability is one of the more troublesome features of ferric hydroxamate methods. It may be caused by excessive amounts of hydroxylamine which must therefore be avoided. Hill (30) removed surplus hydroxylamine with hydrogen peroxide. A similar color stabilizing effect is accomplished by substituting ferric perchlorate for ferric chloride as was done by Nailor et al. (11).

The temperature of the reaction mixture rises when HCl is added, and since color development varies with temperature, it is important that temperatures be kept the same in the standards as in the sera. The use of a constant temperature bath is essential. Stern and Shapiro (17) carried out the reaction at room temperature, Nailor et al. (11), at 30°C. We have adopted 25°C. mainly because a bath maintained at this temperature was available. Because of fading, the number of specimens and standards examined at any time should be limited.

Turbidity occurred infrequently. It was only troublesome in specimens rich in cholesterol in which the higher fatty acid esters of cholesterol formed poorly-soluble products. The use of minimal volumes of serum and of filtrate was helpful. When turbidity appeared, substituting a 1-ml. aliquot of alcohol-ether extract for 3 ml. was almost always effective. The addition of ether, advocated by Stern and Shapiro (11) was less satisfactory.

The ferric hydroxamate color obeys the Beer-Lambert Law over a considerable range of concentrations. It will be noted that the standards used covered a fivefold range. Experiments in which 0.1, 0.2, and 0.3 ml. samples of the same serum were examined gave a linear relationship between sample size and absorbance. Recoveries averaged 100%, with a precision of ±5%. Gey and Schön (15) reports a precision of 3%; Seckfort and Andres (31) 2%.

TEFA IN HEALTHY PERSONS

A study of 78 healthy persons by the method described in this paper gave results in good agreement with those reported by others. Figure 1

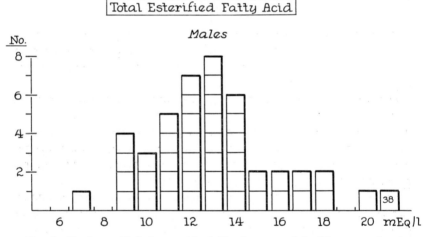

FIG. 1. Total esterified fatty acid of blood serum of healthy men after an overnight fast.

shows the distribution of TEFA concentrations of 44 Caucasoid males ranging in age from 22–60 years. The lowest was 6.8 meq./l., the highest 37.9 meq./l. If the latter is disregarded, 95% of the results fall between 9 and 18 meq./l. with the mode at 13 meq./l. The higher con-

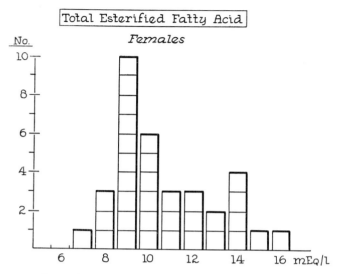

FIG. 2. Total esterified fatty acids of blood serum of healthy women after an overnight fast.

centrations were mainly in the 40–60 year age group. Figure 2 shows TEFA concentrations of 34 women (32 Caucasoids, 2 Negroids). The age range was 17–30 years. Of the values, 95% ranged between 7.6 and 15.0 meq./l.; however, the distribution is much more skewed than that of the males and the mode is only 9 meq./l. Ackermann et al. (32) have discussed the changes in TEFA associated with age. Nailor et al. (11) reported that 42 normal adults in fasting state had serum total esterified fatty acids ranging from 8.6–20.8 meq./l. with a mean of 12.7 meq./l. Peters and Man (33) reported a mean of 12.3 meq./l., with a much wider range, 7.3–36.9 meq./l. Seckfort and Andres (31), found in males a mean of 10.14 and a range of 7.43–17.46 meq./l., and in females a mean of 7.29 and a range of 6.21–11.98 meq./l. Certain subclinical illnesses may result in elevated serum lipid concentrations. For this reason, high values are suspect even when observed in apparently healthy persons. It is for this reason that the highest concentration of TEFA in Fig. 1 was excluded in establishing limits of normal.

The occurrence of occasional extremely high concentrations of TEFA in seemingly healthy persons was previously described by Peters and Man (33). A markedly elevated concentration of triglyceride was responsible.

94   J. G. REINHOLD, V. L. YONAN, AND E. R. GERSHMAN

SIGNIFICANCE OF TEFA

Total esterified fatty acid provides an estimate of the total lipid concentration of plasma since each of the three major lipid types contributes to it. In this respect it may be more informative than separate measurements of cholesterol which do not consistently reflect the total lipid concentration. Neither does the presence of lactescence dependably demonstrate the presence of high lipid concentrations, since the lipid may be dispersed for various reasons such as disproportionate increases in phospholipid and cholesterol. The combination of serum total cholesterol and TEFA measurements is useful in that hypercholesterolemia as well as hyperlipemia will be demonstrated. TEFA, however, should not be regarded as a complete substitute for triglyceride estimations.

TRIGLYCERIDE IN HEALTHY PERSONS

The median triglyceride concentration found in a group of 37 healthy males (mainly professional and business men) by the method described in this paper was 5.2, the mode 5.0, and the mean 5.66 meq./l. The lowest value was 1.5, the highest 27.9 meq./l. (Fig. 3).

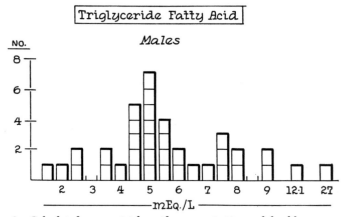

FIG. 3. Calculated serum triglyceride concentrations of healthy men.

Ages ranged from 25 to 55 years. In 20 females, the median was 2.5, the mode 2.0, and the mean 2.1 meq./l. The lowest value was 0.2 meq./l. and the highest was 5.0 meq./l. Ages in the female group, who were mainly laboratory technicians, were considerably lower

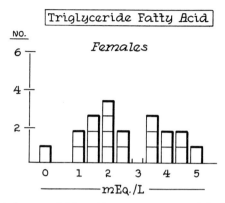

FIG. 4. Calculated serum triglyceride concentrations of healthy women.

than that of the males. The mean of the entire group is 4.56 meq./l. Peters and Man (33), found a mean of 3.1 meq./l. The important influence of age, sex, and diet might account for the difference; however, it appears probable that the method described in this paper yields higher values than the titrimetric method. Extremely high values were excluded when the means were calculated.

SIGNIFICANCE OF TRIGLYCERIDE

Triglyceride frequently varies independently of cholesterol and phospholipid in plasma. In conditions such as idiopathic hyperlipemia, triglyceride may be enormously increased without a proportional rise in cholesterol and phospholipid (34, 35). Although such sera usually are lactescent this is not always true. Gross lactescence appears when triglyceride exceeds 20 meq./l. [Albrink et al. (36)]. In some malnourished persons, Man and Gildea (37) found elevated triglyceride concentrations in conjunction with lowered concentrations of serum cholesterol and phospholipid. The lipemias of diabetes, nephritis, and nephrosis are often characterized by increases in triglyceride concentrations that are much greater than those of other lipids. In hyperthyroidism, Peters and Man (38), found triglyceride concentrations to be elevated in the presence of hypocholesterolemia. In myxedema, also, the two varied independently. The presence of elevated triglyceride has been associated with a tendency to arteriosclerotic disease, (39). Albrink and Man (40) observed that 70% of patients with coronary artery disease had triglyceride concentrations exceeding

5.9 meq./l. Changes in blood clotting also have been related to hyper-lipemia characterized by elevated triglyceride (41). Extreme eleva-tions of lipid, especially triglyceride, have been reported by Albrink and Klatskin (42), Richman (43), and Corazza and Myerson (44) in some patients with abdominal crises associated with pancreatitis. High triglyceride concentrations in serum may occur following over-indulgence in alcoholic beverages.

OTHER METHODS

In addition to those listed in the introduction, two new methods for triglyceride based upon measurement of glycerol after removal of phospholipid have been described by Van Handel and Zilversmit (25) and by Carlson and Wadstrom (45). Blix (46) had used glycerol determinations much earlier as a means of measuring blood lipids. Glycerol is converted to formaldehyde which is determined by use of chromotropic acid. Phospholipids together with the glycerol present in these substances were removed by adsorption on Doucil by Van Handel and Zilversmit (25) and on silicic by Carlson and Wadstrom (45). Comparison of the chromotropic acid method of Van Handel and Zilversmit with the results of the hydroxamate method as de-scribed in this paper showed the results of triglyceride measurements by both methods to be in agreement. Occasional differences in indi-vidual serum specimens were as large as 2 meq./l. In view of the summation of errors involved in calculation of triglyceride from TEFA by correction for phospholipid and cholesterol ester fatty acid; this is not considered excessive. The use of Doucil in the Van Handel and Zilversmit method for the removal of phospholipid also introduces uncertainties which may contribute to the differences. Either method may be used, therefore, with confidence for establishing the presence of elevated triglyceride concentrations in serum.

Kibrick and Skupp (6) found fatty acid concentrations measured in serum by the hydroxamate method to agree with those of their oxidative method.

APPLICATION TO ZONE ELECTROPHORESIS

Tompsett and Tennant (29) have applied the hydroxamate method to measurement of the quantities of esterified fatty acid in the several zones of serum proteins separated by electrophoresis on paper in barbiturate buffer. In our hands, this method for lipoprotein electrophoresis has at times given excellent results, but with unexplained aberrations at other times. These arise partly from the marginal amounts of esterified fatty acids available unless serum fatty acid concentrations are elevated. Mainly they result from translocation of the lipid zones in relation to pilot patterns.

REFERENCES

1. Peters, J. P., and Van Slyke, D. D., "Quantitative Clin. Chem.," Vol. 1, Interpretations, 2nd Ed., Williams & Wilkins, Baltimore, Maryland, 1947.
2. Peters, J. P., and Van Slyke, D. D., "Quantitative Clinical Chemistry," Vol. 2, Chap. IX, Williams & Wilkins, Baltimore, Maryland, 1932.
3. Stewart, C. P., and White, A. C., The estimation of fat in blood. *Biochem. J.* 19, 840 (1925).
4. Stoddard, J. D., and Drury, P. E., A titration method for blood fat. *J. Biol. Chem.* 84, 741 (1929).
5. Bloor, W. R., The determination of small amounts of lipid in blood plasma. *J. Biol. Chem.* 77, 52 (1928).
6. Kibrick, A. C., and Skupp, S. J., Colorimetric method for the determination of fatty acids by oxidation with dichromate. *Arch. Biochem. Biophys.* 44, 134 (1953).
7. Feigl, F., Anger, V., and Frehden, O., II. Detection of carboxylic acids and their derivatives (anhydrides, esters, halides). *Mikrochemie* 15, 9 (1934).
8. Hill, V. T., Colorimetric determination of fatty acid and esters. *Ind. Eng. Chem. Anal. Ed.* 18, 317 (1946).
9. Bauer, F. C., Jr., and Hirsch, E. F., A new method for the colorimetric determination of the total esterified fatty acids in human sera. *Arch. Biochem.* 20, 242 (1949).
10. Stern, I., and Shapiro, B., A new method for colorimetric determination of total esterified fatty acids in human serum. *J. Clin. Pathol.* 6, 158–160 (1953).
11. Nailor, R., Bauer, F. C., Jr., and Hirsch, E. F., Modifications in the hydroxamic acid method for estimation of esterified fatty acids in small amounts of serum. *Arch. Biochem. Biophys.* 54, 201–5 (1955).
12. Kornberg, A., and Pricer, W. E., Jr., Enzymatic synthesis of the coenzyme A derivatives of long chain fatty acids. *J. Biol. Chem.* 204, 329 (1953).
13. Jarrier, M., and Polonovski, J., Méthode de dosage de acides gras estérifiés et son application au sérum sanguin. *Bull. Soc. Chim. Biol.* 37, 495 (1955).
14. Jonnard, R., Rapid colorimetric method for analysis of neutral fats and fatty acids in biological materials. *Clin. Chem.* 2, 254 (1956).
15. Gey, K. F., and H. Schön. Zur Bestimmung von Fettsäureestern im Blut-und Organ-fettem. *Hoppe-Seylers Z. Physiol. Chem.* 305, 149 (1956).
16. Rosenthal, H. L., Pfluke, M. D., Callerami, J., The colorimetric estimation of serum fatty esters. *Clin. Chim. Acta* 4, 329 (1959).
17. Goddu, R. F., Le Blanc, N. F., and Wright, C. M., Spectrophotometric determination of esters and anhydrides by hydroxamic acid reaction. *Anal. Chem.* 27, 1351 (1955).

18. Bayer, E., and Reuther, K. H., Photometrische Mikrobestimming von Acylgruppe analytische Verwendung von Eisen (III) Hydroxamsäurekomplexion I. Mitteilung. *Chem. Ber.* **89**, 2541 (1956).

19. Hestrin, S., The reaction of acetylcholine and other carboxylic acid derivatives with hydroxylamine and it's analytical application. *J. Biol. Chem.* **180**, 249 (1949).

20. Jandorf, B. J., Chemical reactions of nerve gases in neutral solution I. Reactions with hydroxylamine. *J. Am. Chem. Soc.* **78**, 3686 (1956).

21. Schwert, R. S., Determination of amino acid hydroxamides. *Biochim. Biophys. Acta.* **18**, 566 (1955).

22. Goldenberg, V., and Spoerri, P. E., Colorimetric determination of carboxylic acid derivatives as hydroxamic acids. *Anal. Chem.* **30**, 1327 (1958).

23. Wainfan, E., and van Bruggen, J. T., Hydroxylamine and reactive acyls. *Arch. Biochem. Biophys.* **70**, 43 (1957).

24. Rapport, M. M., and Alonzo, N., Photometric determination of fatty acid ester groups in phospholipides. *J. Biol. Chem.* **217**, 193 (1955).

25. Van Handel, E., and Zilversmit, D. B., Micromethod for the direct determination of serum triglycerides. *J. Lab. Clin. Med.* **50**, 152 (1957).

26. Phillips, G. B., Composition of human serum lipoprotein fractions separated by ultrafiltration. *J. Clin. Invest.* **38**, 489 (1959).

27. Kaplan, A., Unpublished experiments cited by Entenman, C. (28).

28. Entenman, C., Preparation and determination of higher fatty acids. In "Methods in Enzymology." Vol. 3, p. 323 (S. P. Colowick and N. O. Kaplan, eds.) Academic Press, New York (1957).

29. Tompsett, S. L., and Tennant, W. S., A method for determining esterified fatty acid with zone electrophoresis of serum proteins. *Am. J. Clin. Pathol.* **26**, 1226 (1956).

30. Hill, U. T., Colorimetric determination of fatty acids and esters. *Anal. Chem.* **19**, 932 (1947).

31. Seckfort, H., and Andres, E., Über den Fettsäuregehalt des Blutserums Gesunder (Mit einem methodischen Erfahrungsbericht) *Deut. Z. Verdauungs Stoffwechselkrank.* **15**, 49 (1956).

32. Ackermann, P. G., Toro, J., Kheim, T., and Kaunitz, W. B., Blood lipids in young and old individuals. *Clin. Chem.* **5**, 100 (1959).

33. Peters, J. P., and Man, E. B., Interrelations of serum lipids in normal persons. *J. Clin. Invest.* **22**, 707 (1943).

34. Holt, L. E., Jr., Aylward, F. X., and Timbres, H. Q., Idiopathic familial hyperlipemia. *Bull. Johns Hopkins Hosp.* **64**, 279 (1940).

35. Holt, L. E., Jr., In "Fat Metabolism: A Symposium on the clinical and Biochemical aspects of Fat Utilization in Health and Disease." (V. A. Najjar, ed.) Johns Hopkins Univ. Press, Baltimore, Maryland, (1954).

36. Albrink, M. J., Man, E. B., and Peters, J. P., The relation of neutral fat to lactescence of serum. *J. Clin. Invest.* **34**, 47 (1955).

37. Man, E. B., and Gildea, E. F., Serum lipoids in malnutrition. *J. Clin. Invest.* **15**, 203 (1936).

38. Peters, J. P., and Man, E. B., Interrelations of serum lipids in patients with thyroid disease. *J. Clin. Invest.* **22**, 715 (1943).

39. Horlick, L., Effect of acute fat loads on serum lipids in atherosclerosis. *Circulation Research* **5**, 368 (1957).
40. Albrink, M. J., and Man, E. B., Serum triglycerides in coronary artery disease. *A. M. A. Arch. Internal Med.* **103**, 4 (1959).
41. Keys, A., Diet and epidemiology of coronary heart disease. *J. Am. Med. Assoc.* **164**, 1912 (1957).
42. Albrink, M. J., and Klatskin, G., Lactescence of serum following episodes of acute alcoholism and it's probable relationship to acute pancreatitis. *Am. J. Med.* **23**, 26 (1957).
43. Richman, A., Acute Pancreatitis. *Am. J. Med.* **21**, 246 (1956).
44. Corazzo, L. J., and Myerson, R. M., Essential hyperlipemia. Report of four cases with special reference to abdominal crises. *Am. J. Med.* **22**, 258 (1957).
45. Carlson, L. A., and Wadstrom L. B., Determination of glycerides in blood serum. *Clin. Chim. Acta* **4**, 197 (1959).
46. Blix, G., Occurrence of triglycerides (neutral fats) in blood plasma. *Biochem. Zeit.* **305**, 145 (1940).
47. Sperry, W. M., In "Methods of Biochemical Analysis," Vol. 2 (D. Glick, ed.) Wiley (Interscience). New York, 1955.

GLUCOSE (ENZYMATIC)*

Submitted by: FRANK W. FALES, Department of Biochemistry and Clinical Research Center, Emory University, Atlanta, Georgia

Checked by: DAVID SELIGSON, Yale University, New Haven, Connecticut

Introduction

The enzymatic method for glucose proposed by Keston (1) in 1956 has received considerable attention both in this country and abroad because of its simplicity, precision, and supposed specificity. Also, its widespread adoption for clinical determinations was stimulated by availability of conveniently packaged reagents.[1] Although simple from a procedural point of view, the method is complicated from a chemical standpoint. The reagent contains the enzymes "glucose oxidase" (more properly classified as dehydrogenase) and peroxidase along with o-dianisidine, a chromogenic hydrogen donor for the peroxidase reaction. Glucose oxidase is quite specific for β-D-glucose (2) although 2-deoxyglucose is oxidized at an appreciable rate whereas mannose and ribose are slowly oxidized (3, 4). Glucose oxidase has flavin-adinine dinucleotide (FAD) as prosthetic group (4). The FAD is initial hydrogen acceptor for the oxidation of glucose to gluconolactone (5) and in turn the reduced FAD is reoxidized by molecular oxygen dissolved in solution, the product of the reaction being hydrogen peroxide. In the presence of excess peroxidase and chromogen, the hydrogen peroxide is rapidly reduced to water and the chromogen oxidized to a colored product. A number of modifications have appeared. Huggett and Nixon (6) and Saifer and Gerstenfeld (7) increased the sensitivity by using a 30-minute incubation period at 37°C. rather than 10 minutes at room temperature; Middleton and Griffiths (8) used the more sensitive but unstable o-tolidine chromogen; Salomon and Johnson (9) and Marks (10) developed a more stable o-tolidine reagent by replacing phosphate with acetate buffer; and Beach and Turner (11) used o-anisdine as chromogen. McComb and

---

* Based on the method of Keston (1).

[1] Glucostat, Worthington Biochemical Corp., Freehold, N. J., and Test Combination, Boehringer & Soehne GmbH, Mannheim, Germany.

Yushok (12) observed increased sensitivity with *o*-dianisidine chromogen by adding excess sulfuric acid at conclusion of incubation which converted the colored product from amber to deep pink and this modification was incorporated into the methods of Kingsley and Getchell (13) and Washko and Rice (14). Also the latter authors stabilized the reagent with added glycerol. As protein precipitation reagent for blood glucose, zinc hydroxide was the usual choice although Saifer and Gerstenfeld (7) used cadmium hydroxide and Kreutzer (15) used perchloric acid.[2] However, Teller (16), Huggett and Nixon (6), Kingsley and Getchell (13), Cawley *et al.* (17), and Raabo and Terkildsen (18) described methods in which serum or plasma was used directly without prior protein precipitation. Thus, a number of modifications of the enzymatic method for glucose have appeared and in consideration of the complexity of the analytical system from a chemical standpoint, it is not surprising that reservations have been expressed (19, 20) concerning the specificity of this method.

The emphasis placed by most authors upon specificity of their method as chief advantage, based on specificity of glucose oxidase, does not seem justified because of many sources of interference. This emphasis upon specificity has engendered the false supposition that various modifications of the method yield true blood glucose values regardless of experimental conditions. As an example, Marks (12) stated that all glucose oxidase methods for blood glucose yield the same normal values whereas the different reducing sugar methods yield a variety of normal values. On the grounds of eliminating this confusion, he recommended the general adoption of glucose oxidase methods. With the appearance of new modifications of the glucose oxidase method, this supposed advantage was not realized, but rather, a new array of normal values was propagated and the problem was compounded rather than eliminated. There exist a number of potential sources of interference for the colorimetric glucose oxidase methods. Obviously, enzyme activators or inhibitors must not be introduced with the sample or reagents. At moderately low pH, fluoride and to a lesser extent, chloride, inhibit peroxidase (21), and this limits the usefulness of the methods employing *o*-tolidine as the chromogen since at the pH required with this chromogen, fluoride cannot be used as preservative (9, 10). Also, unsuitable or improperly balanced protein precipitation reagents may shift the pH from optimal or intro-

---

[2] The use of perchloric acid filtrate may be quite prevalent in Europe because of its use in the procedure outlined with the Boehringer & Soehne Kit.

duce inhibitory amounts of heavy metals. A potential source of positive error are peroxides released by tissues (including whole blood) by acid (9) or peroxides released from reactions on ion-exchange resins (9). An important source of negative error is the presence of reducing agents capable of competing with the chromogen as hydrogen donors for peroxidase or capable of reducing the colored product to colorless form. Uric acid (15, 17), ascorbic acid (15), glutathione (15, 20), cysteine (20), bilirubin (18), thymol (9), and the catechols are reducing agents that interfere, so must not be present in appreciable concentrations. Oxidizing agents capable of converting chromogen to a colored product cause a positive error, i.e., chlorine from tap water contamination (9). Also, the autooxidation by molecular oxygen catalyzed by light may cause error due to exposure of sample and standard to different light intensities (15). Unless a specially purified glucose oxidase preparation free of amylase and maltase activity is used, amylase substrates such as glycogen cause a positive error (20, 22). Thus there is an imposing list of potential sources of error, and hence, despite the specificity of the glucose oxidase, some assurance must be at hand that none of these are operative before the glucose values can be considered as valid.

The modified glucose oxidase procedure and reagent described by Washko and Rice (14) have many advantages, and a modification thereof which is especially adaptable for semiautomatic determination of glucose will be presented. The reagent is quite stable both in storage and use, it is economical in amount of reagent required, and the sensitivity and reproducibility obtained with the reagent are excellent. The reagent cannot be used for direct determination of glucose in serum, plasma, spinal fluid, or urine because these contain interfering materials. However, Somogyi (23) zinc hydroxide filtrates of whole blood, plasma, and spinal fluid are essentially free of interfering materials.

Reagents

1. *Zinc sulfate solution, 2.2%.* Dissolve 22 g. reagent grade $ZnSO_4$·$7H_2O$ in distilled water and bring to volume of 1 l.

2. *Barium hydroxide solution, saturated.* Bring 1 l. distilled water to vigorous boil in 2-l. Erlenmeyer flask, add about 80 g. reagent grade $Ba(OH)_2$·$8H_2O$ from a freshly opened bottle, insert stopper having soda lime trap, mix thoroughly, and set aside for several days to settle. At 25°C., the solution is about 0.45 N.

3. *Barium hydroxide solution, 0.12 N.* Dilute 270 ml. of saturated
barium hydroxide solution to 1 l. with carbon dioxide-free (recently
boiled) distilled water. Barium hydroxide solution in volume of
9.00 ± 0.10 ml. should be required to neutralize 10.00 ml. zinc sulfate
solution. Test by diluting 10.00 ml. zinc sulfate solution with about
25 ml. distilled water, add 2 drops 1% phenolphthalein solution, and
rapidly titrate to faint pink with barium hydroxide solution. Dilute
one or the other solution to proper equivalence. The barium hydroxide
solution can conveniently be dispensed by a buret with side arm for
filling and with both buret and reservoir bottle equipped with soda
lime traps. If buret is equipped with overflow vent, the tube from
the vent should extend into the water. The buret assembly can be
attached to a Seligson (24) automatic pipet for measurement of samples. Plastic rather than rubber tubing should be used in the assembly
because of interfering material which may be derived from rubber.[3]

NOTE: If the method is used primarily for determination of glucose in whole
blood, the following protein precipitation reagents may be preferred: Prepare
2.2% zinc sulfate solution as above; 0.1 N sodium hydroxide solution: 4.0 g.
reagent grade NaOH per 1 l. distilled water; and adjust solutions so that 9.8 to
10.0 ml. of base are required to neutralize 10.00 ml. of zinc sulfate solution.

4. *Stock glucose standard, 1%.* Place 1.000 g. dry, reagent grade
glucose in 100-ml. volumetric flask. Dissolve glucose and bring to volume with 0.25% benzoic acid solution.

5. *Dilute glucose standards.* Prepare standards containing 100, 200,
300, and 400 mg. glucose per 100 ml. by diluting 5.0, 10.0, 15.0, and
20.0 ml. of stock standard to 50.0 ml. with 0.25% benzoic acid solution.

6. *Glycerol-buffer solution, pH 7.0.* Dissolve 3.48 g. oven-dried,
reagent grade Na_2HPO_4 and 2.12 g. oven-dried, reagent grade KH_2PO_4
in 600 ml. distilled water and add 400 ml. reagent grade glycerol. The
solution remains free of growth for months when stored in refrigerator.

7. *Glucose oxidase reagent.* To a clean dry mortar, add 500 mg.
"crude" glucose oxidase, 10.0 mg. horse radish peroxidase, RZ 0.3 or
higher (Reinheitzahl or purity number measured by ratio of absorbancy of solutions at 403 and 275 mμ, respectively), and 2 ml. of
glycerol-buffer solution. Grind thoroughly and with glycerol-buffer
solution wash into a graduated cylinder, bring to a volume of
200 ml., and filter. A Buchner funnel with light suction may be

[3] This precaution provided by Worthington Biochemical Corp. with their
reagent.

used if speed is desired. Completely dissolve 20.0 mg. o-dianisidine in 2.0 ml. absolute methanol by intermittent shaking for several minutes and drain into amber bottle. Add filtered enzyme solution, pouring in initial portion rapidly to prevent precipitation of the chromogen. The reagent loses very little sensitivity in 3 weeks when stored in refrigerator.

NOTE: Some may prefer to prepare the reagent with conveniently packaged "Glucostat."[4] In this case, completely dissolve soluble chromogen (small vial) in 1.0 ml. distilled water and drain into amber bottle. Dissolve contents of second vial and dilute to 100 ml. with glycerol-phosphate buffer. Add solution to the amber bottle, pouring in the initial portion rapidly to prevent precipitation of the chromogen. This reagent is similar to the one described except that it contains half as much glucose oxidase and is therefore somewhat less sensitive.

8. *Sulfuric acid, 6 N.* Add 200 ml. reagent grade sulfuric acid to 1000 ml. distilled water with mixing in cold water bath.

Procedure

Draw up sample (whole blood, serum, plasma, or spinal fluid) into tip of 0.5 ml. automatic pipet[5] equipped with 5-ml. buret and wash into 50-ml. Erlenmeyer flask with 4.5 ml. of 0.12 N barium hydroxide solution (or if preferred, 0.1 N sodium hydroxide solution with whole blood). The buret excursion is 5.00 ml. Swirl solution for 10 seconds to insure complete laking of the blood and without further delay, add slowly while swirling, 5.00 ml. of 2.2% zinc sulfate solution with second buret. Mix thoroughly, making certain that the mixture is uniformly opaque with no streaks of liquid or large clumps of precipitate. Allow to stand for 5 minutes and filter or transfer to centrifuge tube and remove precipitate by centrifugation.

NOTE: Identical results were obtained by glucose oxidase method with filtrates (Whatman No. 1 filter paper) and with supernatant fluid after centrifugation. However, significantly higher values (average about 5 mg. per 100 ml. with 1:20 dilution filtrates) were obtained by Somogyi (25) copper-reduction method with filtrates than with supernatant fluid, presumably due to leaching of reducing material from filter paper. Therefore, centrifugation or washed filter paper should be used when comparing the methods or when determining reducing sugars other than glucose by the combined use of the two methods (20).

Prepare blank and standard filtrate in identical manner as sample filtrate using water and 100 mg. per 100 ml. glucose solution respec-

---

[4] Product of Worthington Biochemical Corp., Freehold, N. J.
[5] Obtainable from Arthur H. Thomas Co., Philadelphia, Pa.

tively as starting materials. Draw up filtrate into tip of 0.2 ml. automatic pipet (24) and wash into spectrophotometer tube with 1.00 ml. of glucose oxidase reagent (brought to room temperature after removal from refrigerator). The buret excursion is 1.20 ml. Immediately place the tube in 37°C. water bath equipped with stirrer or circulating system. Add glucose oxidase reagent to duplicate portions of sample filtrates, to duplicate portions of standard filtrate, and to a single portion of blank filtrate at 30-second intervals. After exactly 30 minutes, remove the tubes one at a time and add 6 N H_2SO_4 contained in flask equipped with 5-ml. dispenser[6] by first tipping back the flask to fill
    [6] Obtainable from California Laboratory Equipment Co., Berkeley, Calif.
the dispenser head and then tipping forward to deliver 5.00 ml. of acid. After 5 minutes but within an hour, determine the absorbances at wavelength 540 mμ against the blank tube.

$$\frac{A_{sample}}{A_{standard}} \times 100 = \text{mg. glucose per 100 ml.}$$

The absorbance is usually linearly related to sample glucose concentration up to 400 mg. per 100 ml., but this should be verified with each new batch of glucose oxidase reagent. If the glucose concentration is too high for accurate measure, the filtrate should be diluted to an appropriate glucose concentration. This semiautomatic method for glucose has been used in our clinical laboratory for almost a year and has proved highly satisfactory.

There are a number of satisfactory modifications of the method. The measurements of volumes may be made with conventional pipets rather than with the semiautomatic devices described. The procedure is readily adaptable for microdetermination of glucose in capillary blood by reducing five-fold the volumes for filtrate preparation (0.9 ml. of 0.1 N NaOH plus 0.1 ml. blood plus 1 ml. 2.2% zinc sulfate). Some may prefer to use 0.1 ml. of 1 : 10 dilution filtrate rather than the prescribed 0.2 ml. of 1 : 20 dilution filtrate. Also, about equal sensitivity is obtained using 0.2 ml. of 1 : 10 dilution filtrate with a 10-minute incubation at 37°C., but this modification has the drawback of a narrow range of linear response.

NOTE: With added glucose in quantities up to about 45 μg., the over-all reaction follows first order kinetics quite perfectly, i.e., the rate of reaction is directly proportional to glucose concentration and after any given period of time, the same fraction of the initial glucose has been oxidized. Thus there is a linear relationship at any given incubation time with glucose up to 45 μg. However, at higher glucose concentrations a rate slower than predicted is observed, and a linear relationship between absorbancy and concentration is not observed at any incubation time.

The determination is satisfactory with a 45 minute incubation period at room temperature rather than 30 minutes at 37°C., but with some loss of sensitivity. In small laboratories where equipment and skilled personnel are limited, the following modification may prove superior. Add 0.500 ml. 1 : 20 dilution zinc hydroxide filtrate to colorimeter tube with volumetric pipet, add 2.00 ml. glucose oxidase reagent, mix, and incubate for exactly 30 minutes at room temperature. Add 5.00 ml. 7.5 N sulfuric acid and mix. Thus there are a number of useful modifications which may be specially suited for various situations but with all glucose oxidase methods, the careful preparation of the zinc hydroxide filtrate is of paramount importance for obtaining satisfactory results because it is at this step that interfering materials are removed.

Discussion

Although the specificity of the colorimetric glucose oxidase methods have been emphasized, few investigators have presented evidence in support of the specificity. Middleton and Griffiths (8), Beach and Turner (11), Kingsley and Getchell (13), and Fales et al. (20), have found agreement between enzymatic and the more specific reducing sugar methods and present this agreement in support of the specificity. On the other hand, Saifer and Gerstenfeld (7) and Campbell and Kronfeld (26) obtained lower values with their glucose oxidase methods and considered this indicative of greater specificity of the glucose oxidase method compared to the reducing sugar method. Also, as has been mentioned previously, there is disagreement concerning normal values. The normal fasting blood sugar values with various modifications of the glucose oxidase method are shown in Table I. The data obtained in this investigation suggest that 60 and 105 mg. per 100 ml. would encompass a great majority of fasting venous blood sugar values of normal individuals. These data are in agreement with those of Kreutzer with zinc hydroxide filtrates. The lower values with perchloric acid filtrates are not unexpected because of interference of glutathione present in these filtrates, but the lower values obtained by Middleton and Griffiths and by Marks (capillary blood) are not readily explained. The normal plasma glucose level was not investigated, but with the assumption of a 45% hematocrit

TABLE I

NORMAL FASTING BLOOD GLUCOSE LEVELS (MG. PER 100 ML.)
REPORTED FOR GLUCOSE OXIDASE METHODS

Number of determinations	Mean	Standard deviation	Range	Filtrate	Reference
			Whole blood		
	(70)			50–90	$Zn(OH)_2$[a]	Middleton and Griffiths (8)
	(70)			50–90	$Zn(OH)_2$[a]	Marks (10)
67	72	11		$HClO_4$	Kreutzer (19)
	82	8		$Zn(OH)_2$	Kreutzer[b]
94	81	9	63–104	$Zn(OH)_2$	Present investigation
			Plasma or serum		
33	91	8		$Zn(OH)_2$	Saifer and Gerstenfeld (7)
54	80	8		None	Raabo and Terkildsen (18)
39	79	10		None	Cawley et al. (17)
13	73		63–88	None	Kingsley and Getchell (13)

[a] Reagents prepared in physiological saline (0.9% NaCl).
[b] Personal Communication.

and with the known corpuscular to plasma ratio of 80 : 100 for glucose (27), the plasma glucose level should average about 10% higher than whole blood level, yielding a mean level of about 89 mg. per 100 ml. This is in agreement with the normal level for plasma found by Saifer and Gerstenfeld using zinc hydroxide filtrates, but it is considerably higher than the levels obtained with methods which use serum or plasma directly without prior protein precipitation. Because of these variable results, it is obvious that at least some of these methods do not yield true blood glucose values, and the question remains, which, if any of these methods do yield true glucose values.

An approach which may be of value in establishing specificity is the recovery study. However, when carried out in the usual way, it has no value in establishing the specificity of the glucose oxidase method as is shown by the following example. Recovery of glucose added to blood proved to be 100% both when zinc hydroxide and tungstic acid filtrates were used, but the blood sugar values with zinc hydroxide filtrates averaged 13 mg. per 100 ml. higher than those with tungstic acid filtrates. Apparently, the glutathione present in tungstic acid filtrate causes a constant underestimation dependent upon the amount present that can be oxidized, and as long as an excess of glucose is present, added glucose is recovered quantitatively. In order to circumvent this difficulty, the endogenous glucose was first removed from the sample by treatment with glucose oxidase; 1 ml. sample was mixed with 1 ml. glucose oxidase solution (10 mg.) in a 125 ml. Erlenmeyer flask to provide adequate aeration. After several hours, a 1 : 10 dilution protein-free filtrate was prepared and 0.1 ml. of filtrate placed in spectrophotometer tube along with 0.1 ml. of glucose solution (10 μg.). The glucose could not be added to the mixture prior to the separation of the protein because the glucose oxidase retains some activity even in the presence of protein precipitation reagents. Blanks and standards were usually prepared containing 0.1 ml. of a filtrate of glucose oxidase solution.

NOTE: An inhibition was found with glucose standards prepared as tungstic acid and copper tungstate filtrates, so standards diluted with water were used in these instances. The preparation of standards and blanks with reagents used for preparation of blood filtrates is especially appropriate in the semiautomatic method described because standard, blank, and test solutions are diluted with the same volumetric equipment, so errors in calibration are completely compensated. However, the contention that these preparations represent reagent controls is not a valid one because the composition of the standard and blank filtrate always differs markedly from the composition of the blood filtrate due to combination of precipitation reagent with protein. Thus, in some instances water more closely approximates the composition of the blood filtrate than do the reagents in the absence of protein. A case in point is the determination of glucose in copper tungstate filtrates of whole blood by the glucose oxidase method. Quite good recoveries were obtained when compared with standards diluted in water, but the "recovery" averaged 115% when compared with standards containing copper tungstate. Therefore a critical evaluation should be carried out to preclude possible interference before incorporation into a procedure of the filtrate method for preparation of blanks and standards. Zinc hydroxide precipitation reagents have no effect upon the glucose oxidase method for glucose, but this may not be true of the reducing sugar methods. The reagents cause a lowering of the blank with copper reduction methods, presumably by inhibition of autoreduction during the heating period. The absorbance of the glucose standard is similarly lowered, but it is not certain that the blood filtrates are identically affected since polarographic measurements indicate that the blank and standard filtrates contain 40% more zinc than do the blood filtrates.

The recoveries of glucose in the presence of various filtrates is shown in Table II. It does not appear that the prior treatment with glucose oxidase altered the interfering materials since poor recoveries were obtained with tungstic acid filtrates of whole blood and with urine filtrates, which are known to contain materials which compete with the chromogen as hydrogen donors for peroxidase. The average

110

TABLE II

RECOVERY OF ADDED GLUCOSE (10 μg.)

Filtrate	Whole blood (%)		Plasma (%)	Spinal fluid (%)	Urine (%)
Zn(OH)₂	97[a]	98[b]	101[a]	101	87[a]
	99	97	99	104	58
	101	104	101	101	86
	98	98	99	102	
	101	98	100	99	
	99	102	98	100	
	98	97	98	98	
Cd(OH)₂	101		100	104	77
	94		98	100	
	99		97	100	
CuWO₄	99[c]		60[d]		
	93		48		
	97		54		
H₂WO₄	83		93		
	84		91		
	84		96		
	90		94		

[a] Prepared with Somogyi (23) zinc sulfate, barium hydroxide reagents.
[b] Prepared with Somogyi (28) zinc sulfate, sodium hydroxide reagents.
[c] Prepared with Somogyi (29) 7% copper sulfate, 10% sodium tungstate reagent for whole blood.
[d] Somogyi's 5% copper sulfate, 6% sodium tungstate for plasma.

recovery with zinc hydroxide filtrates of whole blood was 99.1%; of plasma, 99.1%; and of spinal fluid, 100.9%. Cadmium hydroxide filtrates also appeared satisfactory, as did copper tungstate filtrates of whole blood. However, the copper tungstate reagent recommended for plasma contains too high a concentration of copper. A more rigorous test for traces of interfering materials in zinc hydroxide filtrates of whole blood was carried out by using 0.2 ml. of 1 : 10 dilution filtrate free of glucose plus 0.1 ml. glucose standard (20 μg.) and only a 5 minute incubation period rather than 30 minutes. It has been shown that glutathione exerts its interference early in the course of the reaction (20). With six determinations, the average recovery was 99.0%. This strongly suggests that zinc hydroxide filtrates are essentially free of interfering materials. These data also rule out possibility of fluoride interference since bloods contained fluoride. Comparison of values obtained with plasma (10 μl.) directly without prior protein removal with values obtained with zinc hydroxide filtrates yielded an average recovery of 80%. A similar study with spinal fluid yielded an average recovery of 95%, but when 100 μl. rather than 10 μl. spinal fluid were used, the recovery was only 71%. Several methods (9, 10, 11) have been proposed for preparing urine suitable for determination of glucose by colorimetric glucose oxidase method, but none are reliable for determination of glucose in normal urine. Thus in summary, it would appear that the glucose oxidase method presented yields values which may closely approach the true glucose levels of whole blood, plasma, and spinal fluid provided that carefully prepared zinc hydroxide filtrates are used in the determinations. It also has the advantages of high degree of precision and reproducibility. Under ideal conditions of volume measurements with the same equipment and absorbance measurements with same spectrophotometer tube, fourteen filtrates of glucose standard (10 μg. glucose analyzed) yielded a coefficient of variation of 1.17%. This is at least equal to the precision obtainable by reducing sugar methods with 100 μg. glucose analyzed. Thus the method has the advantages of simplicity of procedure, stable reagent, sensitivity, precision, and accuracy.

REFERENCES

1. Keston, A. S., Specific colorimetric enzymatic reagents for glucose. *Abstract of Papers, 129th Meeting, Am. Chem. Soc.*, p. 31c., April 1958.
2. Keilin, D., and Hartree, E. F., Specificity of glucose oxidase (notatin). *Biochem. J.* **50**, 331–341 (1952).
3. McComb, R. B., Yushok, W. D., and Batt, W. G., 2-Deoxy-D-glucose, a new substrate for glucose oxidase. *J. Franklin Inst.* **263**, 161–165 (1957).
4. Keilin, D., and Hartree, E. F., Properties of glucose oxidase (notatin). *Biochem. J.* **42**, 221–229 (1948).
5. Bentley, R., and Neuberger, A., The mechanism of action of notatin. *Biochem. J.* **45**, 584–590 (1949).
6. Huggett, A., and Nixon, D. A., Use of glucose oxidase, peroxidase, and o-dianisidine in determination of blood and urinary glucose. *Lancet* **2**, 368–370 (1957).
7. Saifer, A., and Gerstenfeld, S., The photometric microdetermination of glucose with glucose oxidase. *J. Lab. Clin. Med.* **51**, 448–460 (1958).
8. Middleton, J. E., and Griffiths, W. J., Rapid colorimetric micro-method for estimating glucose in blood and CSF using glucose oxidase. *Brit. Med. J.* **2**, 1525–1527 (1957).
9. Salomon, L. L., and Johnson, J. E., Enzymatic microdetermination of glucose in blood and urine. *Anal. Chem.* **31**, 453–456 (1959).

10. Marks, V., An improved glucose-oxidase method for determining blood, C. S. F., and urine glucose levels. *Clin. Chim. Acta* 4, 395–400 (1959).
11. Beach, E. E., and Turner, J. J., An enzymatic method for glucose determination in body fluids. *Clin. Chem.* 4, 462–475 (1958).
12. McComb, R. B., and Yushok, W. D., Colorimetric estimation of D-glucose and 2-deoxy-D-glucose with glucose oxidase. *J. Franklin Inst.* 265, 417–422 (1958).
13. Kingsley, G. R., and Getchell, G., Direct ultramicro glucose oxidase method for determination of glucose in biologic fluids. *Clin. Chem.* 6, 466–475 (1960).
14. Washko, M. E., and Rice, E. W., Determination of glucose by an improved enzymatic procedure. *Clin. Chem.* 7, 542–545 (1961).
15. Kreutzer, H. H. Enzymatische bepaling van het bloedsuikergehalte. *Ned. Tijdschr. Geneesk.* 104, 1379–1383 (1960).
16. Teller, J. D., Direct, quantitative colorimetric determination of serum or plasma glucose. *Abstract of Papers, 130th Meeting, Am. Chem. Soc.* p. 69c, Sept. 1956.
17. Cawley, L. P., Spear, F. E., and Kendall, R., Ultramicro chemical analysis of blood glucose with glucose oxidase. *Am. J. Clin. Pathol.* 32, 195–200 (1959).
18. Raabo, E., and Terkildsen, T. C., On the enzymatic determination of blood glucose. *Scand. J. Clin. Lab. Invest.* 12, 402–507 (1960).
19. Kreutzer, H. H., The enzymatic determination of blood-glucose in practice. *1st Colloquium on Enzymes in Clinical Chemistry, Gent (Belgium), April 2, 1960*, Arscia, Brussels, 1960.
20. Fales, F. W., Russell, J. A., and Fain, J. N., Some applications and limitations of the enzymic, reducing (Somogyi), and anthrone methods for estimating sugars. *Clin. Chem.* 7, 389–403 (1961).
21. Chance, B., Peroxidase heme linkages. *Arch. Biochem. Biophys.* 40, 153–164 (1952).
22. Whistler, R. L., Hough, L., and Hylin, J. W., Determination of D-glucose in corn sirup. *Anal. Chem.* 25, 1215–1219 (1953).
23. Somogyi, M., Determination of blood sugars. *J. Biol. Chem.*, 160, 69–73 (1945).
24. Seligson, D., An automatic pipetting device and its applications in the clinical laboratory. *Am. J. Clin. Pathol.* 28, 200–207 (1957).
25. Somogyi, M., Notes on sugar determination. *J. Biol. Chem.* 195, 19–23 (1952).
26. Campbell, L. A., and Kronfeld, D. S., Estimation of low concentrations of plasma glucose using glucose oxidase. *Am. J. Vet. Research* 22, 587–589 (1961).
27. Somogyi, M., Note on the distribution of blood sugar. *J. Biol. Chem.* 90, 731–735 (1931).
28. Somogyi, M., A method for preparation of blood filtrates for the determination of sugar. *J. Biol. Chem.* 86, 655–663 (1930).
29. Somogyi, M., The use of copper and iron salts for deproteinization of blood. *J. Biol. Chem.* 90, 725–729 (1931).

FREE AND CONJUGATED 17-HYDROXYCORTICOSTEROIDS IN URINE*

Submitted by: ROBERT H. SILBER, Merck Institute for Therapeutic Research, Rahway, New Jersey

Checked by: ALFRED M. BONGIOVANNI, Department of Endocrinology, The Children's Hospital of Philadelphia, Philadelphia, Pennsylvania
MATTHEW CARRONA, Grace-New Haven Community Hospital, New Haven, Connecticut
JAMES C. MELBY, Department of Medicine, Boston University School of Medicine, Massachusetts Memorial Hospital, Boston, Massachusetts

Introduction

It is now well established that hypo- or hyperfunction of the adrenal cortex can be detected by determination of urinary steroids possessing a dihydroxyacetone side chain. The fraction of a given dose of hydrocortisone excreted by an individual as 17,21-dihydroxy-20-ketosteroid is reasonably constant so, once determined, it can be used to estimate the total output of his adrenals (6).

In the original "PS" procedure (1), cortisone was extracted from aqueous solution (1 or more μg./ml.) into chloroform, and reagent was added to the dry residue after evaporation of the solvent. Nelson and Samuels (7) adapted the procedure for the determination of hydrocortisone in plasma primarily by introducing a Florosil column separation, the use of the Allen calculation (without blanks), and microcuvettes. Reddy et al. (8) applied the reaction to butanol extracts of urine. This solvent was employed to extract the water soluble steroid glucosiduronates which normally constitute almost 100% of the urinary steroid with the dihydroxyacetone side chain. Glenn and Nelson (9) employed mammalian glucuronidase treatment of urine prior to chloroform extraction and application of the Nelson-Samuels procedure. Silber and Porter (2) omitted chromatographic separation, introduced direct extraction of steroid from chloroform extract into 1/10th volume of reagent, thus avoiding the evaporation step and reducing the blank, and applied bacterial glucuronidase treatment to urine samples. These authors also changed their original reagent

* Based on the method of Silber and Porter (1, 2, 3, 4, 5).

113

which, although satisfactory for cortisone and its tetrahydro derivative (THE), contained too high a concentration of sulfuric acid for
hydrocortisone and its tetrahydro form (THF). In a more detailed
presentation of their procedure (4), these authors proposed a solvent
(carbon tetrachloride) wash and use of a peak (Allen) calculation
after subtraction of the blank readings to improve specificity.

Many additional communications have appeared with procedural
variations far too numerous to discuss here. The analyst is cautioned
regarding the introduction of alterations that are not backed up by
adequate study.

Some of the complications that have been encountered are:

(1) Use of hydrocortisone acetate as the standard—this yields low
standard values due to incomplete extraction from solvent to reagent.

(2) Substitution of mammalian enzyme in Fishman units for bacterial enzyme in sigma units—urines require about ten times as many
Fishman units and a lower pH when mammalian enzyme is used.
(It is advisable to order relatively large amounts of one batch of
enzyme and check in actual practice to make sure that a reasonable
excess of enzyme is being employed.)

(3) Omission or reduction of ethanol content of reagent or increase
in acid content—can cause severe loss of THF.

(4) Excessive exposure to alkali—the side chain is unstable at
alkaline pH.

(5) Excessive time or temperature for color development—can
cause loss of THF.

Principles

One molecule of 17,21-dihydroxy-20-ketosteroid rearranges (Mattox
Reaction) (10) in the acidic reagent to form a 21-aldehyde which
then reacts with one molecule of phenylhydrazine to yield a yellow
phenylhydrazone (2, 4, 11). At room temperature (25°C.) the reaction reaches a maximum in 14–16 hours with a peak at 410 mμ.
A second molecule of phenylhydrazine also reacts at the 3 position
of hydrocortisone and other steroids but this reaction is of little conse-
quence because it yields a lesser peak at about 350 mμ and little of
such steroid is present in urine.

Since reaction with phenylhydrazine to form a yellow product is
obviously not specific for steroids, the specificity of the procedure is
largely dependent upon use of an appropriate solvent for extraction.

Chloroform and methylene chloride both are satisfactory, but the latter is generally preferred.[1]

The use of a sample blank prepared without phenylhydrazine provides a correction for methylene chloride soluble substances that react with the acid-alcohol alone. The blank is also reduced by washing the solvent extract with 0.1 N NaOH and by washing the original aqueous sample with carbon tetrachloride, which extracts very little of the usual[2] urinary 17,21-dihydroxy-20-ketosteroids. Substances that react with the complete reagent but with a different peak are at least partially eliminated by use of the Allen correction (A at 410 mμ times 2, minus the sum of A at 380 mμ and 440 mμ).

Reagents

1. *Methylene chloride.* Merck reagent—spectrophotometric grade.
2. *Carbon tetrachloride.* Merck reagent—spectrophotometric grade.
3. *Absolute ethanol.* U. S. Industrial Chemicals.
4. *Phenylhydrazine hydrochloride, recrystallized from hot ethanol, dried over CaCl$_2$ in vacuo.*
5. *Sulfuric acid, 24 N.* 310 ml. reagent sulfuric acid and 190 ml. distilled water.
6. *Blank reagent.* 100 ml. of 24 N sulfuric acid mixed with 50 ml. absolute ethanol.
7. *PS reagent.* 65 mg. phenylhydrazine hydrochloride in 150 ml. blank reagent.
8. *β-Glucuronidase.* Type 1, Sigma Chemical, St. Louis, Missouri.
9. *Phosphate buffer, pH 6.5.* 60.6 g. Na$_2$HPO$_4$ and 123.4 g. KH$_2$PO$_4$.
10. *Sodium hydroxide, 0.1 N.*

Hydrocortisone standard, 20 μg./ml. 20 mg. hydrocortisone (21-alcohol) in 5 ml. absolute ethanol diluted to 1 l. with water.

Materials and Apparatus

1. Centrifuge tubes,[3] 45 ml., with plastic caps.
2. Test tubes,[3] 150 mm. in length, with 15 mm. inside diameter, fitted with plastic caps.

[1] Use of this solvent first reported by Bongiovanni and Eberlein (12).

[2] Under certain conditions substance S is produced by the adrenal cortex (17). This steroid and its tetrahydro derivative are extracted by both carbon tetrachloride and methylene chloride (18).

[3] Rinsed with absolute alcohol and methylene chloride and allowed to drain dry before use.

3. Plastic caps,[3] No. 590, Lumelite Corporation, Pawling, New York.

4. Beckman DU spectrophotometer.

5. Microcuvettes, $50 \times 12 \times 12$ mm., 0.8 ml. capacity (Pyrocell Manufacturing Company, New York City) with diaphragm attachment.

6. Blunt tip needles, 16 gauge, 6 inches in length.

Collection and Storage of Samples

Samples should preferably be adjusted to pH 6.5 and stored in a deep freeze. It is important to avoid alkalinity because the side chain is less stable at high pH. Various preservatives have been employed with reasonable success, such as 100 mg. phenol or 1 g. boric acid per 100 ml., or a layer of toluene.

Although accurate 24-hour urines are desirable, one seldom can be certain of the timing. Therefore the author prefers to collect 24-hour specimens but to express the steroid excretion on the basis of urinary creatinine.

Procedure

Free steroid—about 5% of the total

(a) Eight milliliters of urine at pH 6.5 is washed twice with 3 volumes of carbon tetrachloride by shaking 20–30 seconds in a 45 ml. centrifuge tube equipped with a plastic cap. After brief centrifugation, the solvent washes are discarded.

(b) Five milliliters of washed urine is shaken 20–30 seconds in a second centrifuge tube with 25 ml. methylene chloride. After centrifugation, the aqueous phase is discarded by aspiration.

(c) Two milliliters of 0.1 N sodium hydroxide is added, and after shaking 10 seconds and centrifuging, the alkaline wash is discarded by aspiration.

(d) Ten-milliliter aliquots of washed methylene chloride extract are placed in each of two test tubes equipped with plastic caps, in one tube with 1 ml. of PS reagent, in the second with 1 ml. blank reagent.

(e) After capping tightly, the tubes are shaken vigorously but carefully (to avoid spray of acid reagent) for 20 seconds. The cap is carefully removed and after brief centrifugation, the solvent is removed by aspiration with a blunt needle, with as little loss of the reagent as possible.

(f) After standing at 25°C. 14–16 hours,[4] the samples are placed in microcuvettes and optical densities are determined at 380, 410, and 440 mμ in comparison with the appropriate blanks from the "zero standard."

(g) Hydrocortisone standards (0–20 μg.) are diluted to 5.0 ml. with water and are carried through the same procedure.

CONJUGATED STEROID

(a) Urine samples at pH 6.5 are washed three times with 3 volumes of methylene chloride and 1.5 to 6 ml. samples are incubated overnight at 37°C., after addition of 250 sigma units of β-glucuronidase per milliliter and 20 mg. of dry phosphate buffer per milliliter. The enzyme and buffer may be added in dry form or in an equal volume of water.

(b) The hydrolyzed samples are then washed twice with 3 volumes of carbon tetrachloride and 1.0 to 5.0 ml. portions are diluted to 5.0 ml. with water and extracted etc., as above (b–g) for free steroid. (If the enzyme and buffer were added in aqueous solution the total volume of samples and standards is 10.0 ml.)

Typical Analyses and Calculations

	380 mμ	410 mμ	440 mμ
Free			
5 ml.	0.054−0.015=0.039	0.048−0.010=0.038	0.034−0.010=0.024
Conjugate			
2 ml.	0.127−0.041=0.086	0.158−0.023=0.135	0.114−0.023=0.091
Standards			
4 μg.	0.053−0 =0.053	0.074+0.001=0.075	0.051−0 =0.051
16 μg.	0.238+0.001=0.239	0.313−0 =0.313	0.216−0.002=0.214

Corrected Optical Densities

Free	$0.038 \times 2 - 0.039 - 0.024 = 0.013$
Conjugate	$0.135 \times 2 - 0.086 - 0.091 = 0.093$
Standard, 4 μg.	$0.074 \times 2 - 0.053 - 0.051 = 0.044$
Standard, 16 μg.	$0.313 \times 2 - 0.239 - 0.214 = 0.173$

$$\frac{4}{0.044} = 91, \quad \frac{16}{0.173} = 92.5 \quad \text{Average factor} = 92$$

[4] If results are urgently needed the samples may be placed in a 60°C. water bath for 30 minutes for color development.

Free $0.013 \times 92 = 1.2$ μg./5 ml.
Conjugate $0.093 \times 92 = 8.55$ μg./2 ml.

If the analyst does not require the precision of the conjugated steroid analysis, he can determine total steroid by following the procedure for conjugated steroid but omitting the methylene chloride washing of the urine.

If microcuvettes are unavailable, conjugated or total steroid may be determined by analyzing 5–10 ml. of urine and substituting 3 or 4 ml. of PS reagent for the usual 1 ml. One also would need to employ higher standards (15–80 μg.) in the same aqueous volume as the urine sample.

Discussion

As in other procedures employing a single standard for the determination of two or more related compounds, clinically useful values can be obtained, even though they may not be absolute. In the procedure described, THE gives readings about 120% those of hydrocortisone, whereas THF gives lower readings, about 70%. If desired, THF can be separately determined (4) and a correction applied. With samples containing normal or high concentrations, a procedure utilizing both THE and THF standards and a relatively simple calculation can yield more quantitative values (4). However, for most purposes the procedure described in this text is adequate.

In the estimation of adrenal inhibition by measurement of urinary 17,21-dihydroxy-20-ketosteroids (13) one must allow for the excretion of 20–40% of the exogenous steroid administered, unless, of course (a), it does not have the typical dihydroxyacetone side chain or (b) it has a 16-hydroxyl group.

The Reddy procedure (8), which does not require enzymatic treatment of the urine, has been very useful clinically, but as Reddy has reported (14) it has the disadvantages of high blanks and low recovery of hydrocortisone (78%), THE (87%), and THF (35%). Cortisone is employed as the standard.

The Norymberski procedure (15, 16) for the determination of "ketogenic steroids" involves the conversion of hydrocortisone and certain related steroids to 17-ketosteroids which are then determined by means of the Zimmerman reaction. This technique also avoids the need for enzyme treatment and is preferred in some laboratories. Since it is not specific for the dihydroxyacetone side chain, it does not yield as low values for the urines of Addisonians and much higher concentrations are found in the urines of patients with adrenogenital syndrome (due to excretion of an abnormal amount of steroids with a partially reduced side chain). Since it is not as specific as the PS procedure it should always yield higher values.

After administration of Metopirone (SU4885-Ciba), which inhibits 11 hydroxylation, large amounts of substance S or 11-deoxyhydrocortisone are produced (17) and the tetrahydro form appears in the urine. These steroids are also determined by the procedure described in this chapter but they can be more specifically determind by employing carbon tetrachloride as the extracting solvent in place of methylene chloride (18).

Although the use of methylene chloride or chloroform for extraction in ordinary use results in a relatively high degree of specificity, it is possible for certain medications to complicate the analyses. For example, quinine, colchicine, and paraldehyde are not only extracted by the solvent but also react with the reagent (3). No doubt there are and will be others. Some can be eliminated by the carbon tetrachloride wash but ideally one should attempt collections during drug-free periods to avoid such difficulties.

Normal Values

In ten children of both sexes, ages 8–12 years, a range of total corticosteroid excretion in mg./g. creatinine was 4.0–8.8, with an average of 5.5 ± 0.5. In 27–49-year-old male adults 3.4–13.4 mg./g. creatinine, with an average of 6.4 ± 0.5 was found.

REFERENCES

1. Porter, C. C., and Silber, R. H., A quantitative color reaction for cortisone and related 17,21-dihydroxy-20-ketosteroids. *J. Biol. Chem.* **185**, 201-207 (1950).
2. Silber, R. H., and Porter, C. C., The determination of 17,21-dihydroxy-20-ketosteroids in urine and plasma. *J. Biol. Chem.* **210**, 923-932 (1954).
3. Silber, R. H., and Busch, R. D., The specificity of the reaction of phenylhydrazine with 17,21-dihydroxy-20-ketosteroids. *J. Clin. Endocrinol.* **15**, 505-507 (1955).
4. Silber, R. H., and Porter, C. C., Determination of 17,21-dihydroxy-20-ketosteroids in urine and plasma. *In* "Methods of Biochemical Analysis" (D. Glick, ed.) pp. 139-169. Wiley (Interscience), New York, 1957.
5. Silber, R. H., The validity of hormone assays applicable to clinical medicine.

1. Section: Corticosteroids. Symposium at International Endocrine Meeting in Copenhagen, 1960.

6. Silber, R. H., Estimation of hydrocortisone secretion. *Clin. Chem.* **1**, 234-240 (1955).

7. Nelson, D. H., and Samuels, L. T., A method for the determination of 17-hydroxycorticosteroids in blood: 17-hydroxycorticosterone in the peripheral circulation. *J. Clin. Endocrinol.* **12**, 519-526, (1952).

8. Reddy, W. J., Jenkins, D., and Thorn, G. W., Estimation of 17-hydroxy-corticoids in urine. *Metabolism* **1**, 511-527 (1952).

9. Glenn, E. M., and Nelson, D. H., Chemical method for the determination of 17-hydroxycorticosteroids and 17 ketosteroids in urine following hydrolysis with β-glucuronidase. *J. Clin. Endocrinol.* **13**, 911-921 (1953).

10. Mattox, V. R., Steroids derived from bile acids XV—the formation of glyoxal sidechain C^{17} from steroids with dihydroxy acetone and Δ^{16} keto sidechain. *J. Am. Chem. Soc.* **74**, 4340-4347 (1952).

11. Barton, D. H. R., McMorris, T. C., and Segovia R., The nature of the Porter-Silber reaction. *J. Chem. Soc.* May, 2027-2031 (1961).

12. Bongiovanni, A. M., and Eberlein, W. R., Determination, recovery, identification and renal clearance of conjugated adrenal corticoids in human peripheral blood. *Proc. Soc. Exptl. Biol. Med.* **89**, 281-285 (1955).

13. Liddle, G. W., Tests of pituitary–adrenal suppressibility in the diagnosis of Cushing's syndrome. *J. Clin. Endocrinol.* **20**, 1539-1560 (1960).

14. Reddy, W. J., Modification of the Reddy-Jenkins-Thorn method for the estimation of 17-OH-corticosteroids in urine. *Metabolism* **3**, 489-492 (1954).

15. Norymberski, J. K., Stubbs, R. D., and West, H. F., Assessment of adreno-cortical activity by assay of 17-ketogenic steroids in urine. *Lancet* **1**, 1276-1281 (1953).

16. Appleby, J. K., Gibson, G., Norymberski, J. K., and Stubbs, R. D., Indirect analysis of corticosteroids. 1. The determination of 17-hydroxy-corticosteroids. *Biochem. J.* **60**, 453-460 (1955).

17. Liddle, G. W., Estep, H. F., Kendall, J. W. Jr., Williams, W. C. Jr., and Townes, A. W., Clinical application of a new test of pituitary reserve. *J. Clin. Endocrinol.* **19**, 875-894 (1959).

18. Henke, W. J., Doe, R. P., and Jacobson, M. E., A test of pituitary reserve utilizing intravenous SU-4885, with a new method for extraction of 11-desoxy-corticosteroids. *J. Clin. Endocrinol.* **20**, 1527-1531 (1960).

THE DETERMINATION OF 5-HYDROXYINDOLEACETIC ACID IN URINE*

Submitted by: HERBERT WEISSBACH, Laboratory of Clinical Biochemistry, National Heart Institute, National Institutes of Health, Public Health Service, U.S. Department of Health, Education, and Welfare, Bethesda, Maryland

Checked by: M. H. WISEMAN-DISTLER, Allan Memorial Institute, Montreal, Quebec, Canada

HARRIET T. SELIGSON, Yale University, New Haven, Connecticut

SEYMOUR WINSTEN, Albert Einstein Medical Center, Philadelphia, Pennsylvania

Although serotonin is present in many tissues, only trace amounts are normally excreted in the urine. However, the enzymatic breakdown of serotonin results in the formation of 5-hydroxyindoleacetic acid (5-HIAA), and this acid is found in milligram quantities in the urine. Since it is the major metabolite of serotonin it is a sensitive index of serotonin metabolism in the body. It has recently been shown that patients with malignant carcinoid excrete large amounts of 5-HIAA in the urine due to the excessive formation of serotonin by the carcinoid tumor (1, 2, 3). A wide survey of various clinical conditions showed that only in these patients does one find an elevated excretion of 5-HIAA (4), and therefore the urinary assay of 5-HIAA provides a simple diagnostic test for this disease.

Principle

The assay for 5-HIAA is based on the color reaction with 1-nitroso-2-naphthol (5). Where the concentration of 5-HIAA is greater than 30 μg per ml., as is true in the urine of most patients with malignant carcinoid, the colorimetric reaction can be employed directly on the urine (6). However, in normal urines the 24-hour urinary excretion of this acid ranges between 2–9 mg. and an extraction procedure is employed both to concentrate the 5-HIAA and to separate it from interfering substances before the color is developed. The 5-HIAA is readily extracted from acid solution into ether, and can be re-extracted

* Based on the methods of Udenfriend *et al.* (5) and Udenfriend *et al.* (7).

back into a small volume of pH 7.0 buffer (7). Large amounts of keto acids interfere with the assay, as does indoleacetic acid. The former are removed by treatment of the urine with 2,4-dinitrophenylhydrazine and the latter by a preliminary chloroform extraction (7).

Reagents

1. 1-Nitroso-2-naphthol. Prepare a 0.1% solution in ethanol.

2. Nitrous acid reagent. To 5 ml. of 2 N H_2SO_4 is added 0.2 ml. of 2.5% sodium nitrite. This reagent is prepared fresh daily.

3. Ethyl ether, reagent grade. Wash once with a dilute solution of ferrous sulfate to destroy peroxides, and then twice with water.

4. 2,4-Dinitrophenylhydrazine (0.5%) in 2 N HCl.

5. 0.5 M Phosphate buffer (0.5 M), pH 7.0.

6. 5-Hydroxyindoleacetic acid, cyclohexylamine salt.[1] A stock solution of 5-HIAA (100 γ/ml.) is relatively unstable and should be freshly prepared each week. It should be kept cold and in dark bottles.

Procedure

1. *Simple Diagnostic Test for Malignant Carcinoid:* To a glass-stoppered centrifuge tube containing 0.8 ml. of water and 0.5 ml. of 1-nitroso-2-naphthol add 0.2 ml. of urine and mix. Add 0.5 ml. of nitrous acid reagent and mix again. Let stand at room temperature for 5–10 minutes then add 5 ml. of ethylene dichloride and shake once more. The phases should be separated by centrifugation. A positive test is indicated by a purple color in the top layer. With normal urine a pale yellow color is observed. Since certain drugs are known to interfere with this color reaction (see below) it is advisable to run an internal standard in which about 10 μg. of 5-HIAA are added to another aliquot of urine and the color developed. Assuming an average 24-hour volume of 1000 ml. a purple color will be seen at levels of 5-HIAA excretion as low as 30 mg. per 24 hours.

2. *Quantitative Extraction Procedure for 5-HIAA:* To 6 ml. of urine in a 50-ml. glass-stoppered bottle are added 6 ml. of 2,4-dinitrophenyl-hydrazine. When large amounts of keto acids are present, as evidenced by visible precipitation, 15–30 minutes should be allowed in order to obtain complete reaction. In the absence of a visible precipitate 10 minutes are sufficient to insure completeness of reaction. Following

[1] Obtained from the California Corp. for Biochemical Research.

this, 25 ml. of chloroform are added and the bottle is shaken for a few minutes and centrifuged. The organic layer is removed by aspiration and replaced with a fresh 25-ml. aliquot of chloroform. After centrifuging a 10-ml. aliquot of the aqueous phase is transferred to a 40-ml. glass-stoppered centrifuge tube containing 4 g. of NaCl and 25 ml. of ether. The tube is shaken for 5 minutes and centrifuged. A 20-ml. aliquot of the ether is then transferred to another 40-ml. glass-stoppered centrifuge tube containing 1.5 ml. of 0.5 M phosphate buffer, 7.0. The tube is shaken for 5 minutes, centrifuged, and the ether layer removed by aspiration. One ml. of the buffer phase is transferred to a 12-ml. glass-stoppered centrifuge tube containing 0.5 ml. of nitroso-naphthol reagent. After mixing, 0.5 ml. of nitrous acid reagent is added, and the sample is mixed again and warmed at 37°C. for 5 minutes. To remove excess nitroso-naphthol and trace amounts of urinary pigments that may come through the extraction procedure 5 ml. of ethyl acetate are added and the tube is shaken. After separation of the phases the aqueous layer is transferred to a 1-ml. cuvette and the optical density is measured at 540 mμ. Standards (10–200 μg.), as well as a reagent blank, are treated exactly as the urine, substituting water for the urine aliquot. Recoveries of known amounts of 5-HIAA added to urine average about 85% of the control standards, and urinary values are corrected for this low value.

NOTE: Checkers M. H. W.-D and S. W. report that better "normal values," that is low values are obtained when an internal standard is added to each sample before analysis.

Discussion and Results

The direct color reaction is a measure of total 5-hydroxyindoles and is not specific for 5-HIAA. However, even in carcinoid urine 80–90% of the hydroxyindole material present is 5-HIAA (2). Since unextracted urine normally contains substances which interfere with the color reaction it cannot be used for precise quantitative data. Certain drugs such as chloropromazine and its metabolites cause inhibition of color formation. Therefore, internal recoveries (5-HIAA added to urine) should always be employed as controls when assaying for 5-HIAA either by the simple direct reaction or the quantitative extraction procedure. If a positive test is observed using the direct color reaction, the extraction procedure should be employed to verify that the reacting material is 5-HIAA. p-Hydroxyacetanilide, in large

124 H. WEISSBACH

amounts, will give a false positive result but will not come through the extraction procedure for the quantitative determination of 5-HIAA. The urine of patients who have consumed cough remedies containing guaiacol have also been shown to yield positive results in this assay. However, the colored products can be distinguished from the nitroso-naphthol-5-HIAA chromophore.

Except for the occasional occurrence of trace amounts of a second acidic 5-hydroxyindole substance, as detected by paper chromatography, extracts of urine, prepared as described in the assay procedure, contain no other nitroso-naphthol reacting material. The normal urinary excretion in man is between 2 to 9 mg. per day. Although there are large individual variations the urinary 5-HIAA excretion in the same individual is relatively constant from day to day. In patients with malignant carcinoid 5-HIAA excretions range from 30–1000 mg. per day.

1. Lembeck, F., 5-Hydroxytryptamine in a carcinoid tumour, *Nature* **172**, 910–911 (1953).
2. Sjoerdsma, A., Weissbach, H., and Udenfriend, S., A clinical, physiological and biochemical study of patients with malignant carcinoid (Argentaffinoma). *Am. J. Med.* **20**, 520–531 (1956).
3. Sjoerdsma, A., Weissbach, H., Terry, L. L., and Udenfriend, S., Further observations on malignant carcinoid. *Am. J. Med.* **23**, 5–15 (1957).
4. Haverback, B. J., Sjoerdsma, A., and Terry, L. L., Urinary excretion of the serotonin metabolite, 5-hydroxyindoleacetic acid, in various clinical conditions. *New Engl. J. Med.* **255**, 270–272 (1956).
5. Udenfriend, S., Weissbach, H., and Clark, C. T., The estimation of 5-hydroxytryptamine (Serotonin) in biological tissues. *J. Biol. Chem.* **215**, 337–344 (1955).
6. Sjoerdsma, A., Weissbach, H., and Udenfriend, S., Simple test for diagnosis of metastatic carcinoid (Argentaffinoma). *J. Am. Med. Assoc.* **159**, 397 (1955).
7. Udenfriend, S., Titus, E., and Weissbach, H., The identification of 5-hydroxy-3-indoleacetic acid in normal urine and a method for its assay. *J. Biol. Chem.* **216**, 499–505 (1955).

DETERMINATION OF PROTEIN-BOUND IODINE IN SERUM*

Submitted by: O. P. Foss†, Medical Research Center, Brookhaven National Laboratory, Upton, New York

Checked by: ALEX KAPLAN, School of Medicine, University of Washington, Seattle, Washington,

R. S. MELVILLE, Veterans Administration Hospital, Iowa City, Iowa

Introduction

The thyroid produces iodine-containing hormones which after entrance into the bloodstream are associated with serum proteins. The iodine moiety of the circulating hormones is called serum protein-bound iodine, PBI (in some publications called serum precipitable iodine, SPI). It is found that a correlation exists between the activity of the thyroid and the level of PBI. Hypofunction is associated with PBI values less than 3.5 µg. per 100 ml. of serum. In euthyroid persons the PBI ranges from 3.5 to 8.0 µg. per 100 ml. of serum, while hyperfunction results in values above 8 µg. per 100 ml. of serum. Knowledge of the PBI level is therefore of importance for the clinicians in many cases.

Several investigators have contributed to the development of the present procedure for PBI determination. Sandell and Kolthoff (1) published in 1937 a new method for determination of iodine in amounts of hundredths of a microgram. In this method, which is used for all PBI determination at the present time, the iodine is measured by means of the catalytic effect it exerts in the reaction between ceric ions and arsenite. Protein-bound iodine has to be transformed into inorganic iodide before the measurement can be done. Trevorrow (2) demonstrated in 1939 that the iodine of the blood exists in two forms: inorganic iodide and organically bound iodine. The latter fraction was found to be associated with proteins. It stayed with the proteins during dialysis and precipitation. In 1942 Riggs et al. (3) showed that the

* Based on the method of Foss et al. (8).
† This research was supported by the U.S. Atomic Energy Commission and The U.S. Public Health Service. The present address of Dr. Foss is: Norwegian Radium-hospital, Oslo, Norway.

125

protein-bound fraction of iodine was associated with the serum protein and not with the erythrocytes. Therefore, determination of protein-bound iodine has to be carried out with serum and not with whole blood, and the inorganic iodide has to be removed.

The first determination of serum protein-bound iodine which fulfilled these requirements was done in 1942 by Riggs et al. (3). The inorganic iodide was removed by dialysis. The protein-bound iodine was transformed into inorganic iodine by wet acid combustion.

In 1944 Salter and McKay (4) introduced the alkaline ashing procedure for the determination of PBI. The precipitated proteins were burned and the PBI was transformed into inorganic iodide by incineration in the presence of sodium carbonate. Salter and McKay precipitated the proteins by acidifying to pH 6.0 and heating. In 1950 Barker and Humphrey (5) modified Salter and McKay's procedure chiefly by substituting Somogyi's zinc hydroxide precipitation of the proteins in place of the acid coagulation. In 1955 Grossmann and Grossmann (6) modified the technique by avoiding excess acid in dissolving the ash. Avoidance of violent evolution of CO_2 appeared to diminish loss of iodine. In 1958 Vilkki (7) introduced potassium hydroxide as alkali for the incineration.

The alkali incineration method has recently been reexamined by Foss et al. (8) in order to ascertain conditions of incineration, resolution of the ash, nature and concentrations of reagents, and conditions for final photometric determination that would yield results least liable to error. The procedure here described is based on the work of previous authors, quoted above, and modified in details as the result of the reexamination (8).

The determination of PBI is an ultramicro procedure in which several factors can interfere. Reliable results can be obtained only when the chosen procedure is strictly followed. Modifications must not be introduced unless they have been proved not to change the results. The precision of the determination depends much upon the skill, experience, and conscientiousness of the analyst.

Principles

The determination of PBI consists of three steps:
1. The removal of the inorganic iodide.
2. The transformation of PBI into inorganic iodide.
3. The determination of the iodide formed from PBI.

In the present procedure the inorganic iodide is removed by precipitating the proteins and washing the precipitate. The PBI is transformed to inorganic iodide by incineration of the precipitated proteins in the resence of potassium hydroxide. The determination of iodine is done in an aliquot part of an extract of the ash by measuring its catalytic effect on the reduction of Ce^{IV} to Ce^{III} by As^{III}. The rate of reduction is followed by the rate of disappearance of the yellow color of the ceric ion.

Equipment and Apparatus

Separate room for the PBI analysis. It is recommended that the analysis be carried out in a separate room away from iodine, mercury, and silver compounds which can cause serious error in results. All glassware should be washed separately from glassware used for other purposes.

Muffle oven and drying oven. These shall be used exclusively for PBI analysis. If the muffle oven can be adjusted so that the temperature can be constant at 115°C, it can also serve as drying oven.

NOTE: It is important that the door and ventilation hole of the muffle oven can be tightly closed when desired.

Incineration tubes. Borosilicate test tubes, 15 by 125 mm. outside measure, 12.5 mm. bore.

NOTE: The tubes become heavily etched during the incineration and cannot be used repeatedly. The amount of iodine adsorbed to the etched surface and not dissolved in the extract of the ash is negligible the first time the tubes are used. The third time they are used, however, about 5% of the iodine was found to be lost by adsorption to the glass and after eight incinerations 7–13% of the iodine was lost in this way. Checker A. K. preferred to discard tubes after one use.

Steel rack for the incineration tubes. The rack ought to be so designed that it can be placed both in an upright position and a horizontal position. In the upright position it keeps the tubes in a nearly vertical position well suited for precipitation and washing of the precipitate (Fig. 1). When it is turned over the tubes stand in a slightly slanted position, about 20° from the horizontal. This position is favorable for the drying and incineration.

Photometer. Any good photometer can be used. It is an advantage that cylindrical cuvettes of about 15 mm. diameter fit into the photometer. If a filter-photometer is used a filter giving maximal transmittance at about 420 mµ. should be chosen.

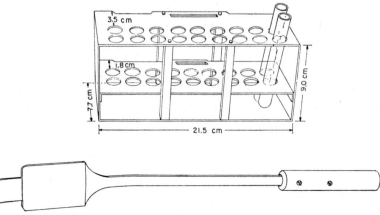

FIG. 1. Rack for the incineration tubes with spade to remove the rack from the hot oven. Published by courtesy of the inventors, E. Nielsen and E. Bjerring, of the Medicinsk Laboratorium, Vognmagersgade 7, Copenhagen.

Cuvettes for the photometric analyses. Ordinary test tubes can be used as cuvettes when they are selected by measuring in them the optical density of the 0.020 N ceric sulfate solution diluted 20 times with 1.6 N sulfuric acid. Tubes showing a difference of more than 1% from the average optical density are rejected.

Footed glass rods. Their length is about 17 cm. with a diameter 3 to 4 mm. One end is spread into a mushroom foot of about 8 mm. diameter.

Reagents

Every time a new batch of a reagent is prepared it should be compared with the remaining part of the old batch by analyses of samples from the same pool of serum in order to secure conformity and avoid contaminations.

Water. The water used for the preparation of the reagents and in the analysis shall either be triply distilled in glass or distilled water passed through a column of ion exchange resin or ion-exchanged water distilled.

NOTE: Contamination with organic matter may interfere with the analysis. Therefore, rubber tubing should be avoided, and the water should not be stored for more than 1 week, as growth of microorganisms occurs even in distilled water.

Zinc sulfate, 10%. 100.0 g. $ZnSO_4$ 7 H_2O per 1000 ml.
Sodium hydroxide, 0.50 N.

NOTE: These reagents are described by Somogyi (9). They should be so related that when 10.00 ml. of the zinc sulfate solution is diluted with 50–70 ml. of water and titrated with the sodium hydroxide solution, 10.8–11.2 ml. of the sodium hydroxide are required to produce a faint permanent pink color with phenolphthalein.

Potassium hydroxide, 2.0 N. Dissolve 70 g. KOH and dilute to 500 ml. Titrate with 0.10 N HCl, calculate and add the amount of water required to get a final strength of 2.0 N.

Sulfuric acid, 7.0 N. Dissolve 98 ml. of concentrated H_2SO_4 in 300 ml. of water. Cool and dilute to 500 ml.

Sufuric acid, 7.0 N, containing HCl, 0.65 N. Dissolve 98 ml. concentrated H_2SO_4 in 300 ml. of water. Cool, add 27 ml. of concentrated HCl and dilute to 500 ml.

Sodium arsenite, 0.100 N. Dissolve 6.50 g. $NaAsO_2$ and dilute to 1000 ml.

NOTE: Checker A. K. preferred 4.95 g. of AS_2O_3 instead of 6.5 g. of $NaAsO_2$.

Ceric sulfate, 0.020 N in 1.6 N sulfuric acid. 12.65 g. of ceric ammonium sulfate, $Ce(SO_4)_2 \cdot 2(NH_4)_2SO_4 \cdot 2 H_2O$, is dissolved in 500 ml. of water plus 230 ml. of 7.0 N H_2SO_4. The mixture is boiled for 5 minutes, cooled and diluted to 1000 ml. It is stored in a dark bottle and can be kept for a month.

NOTE: The author has noticed a turbidity in the final step of the analysis when the solutions of ceric sulfate used had been stored for more than 1 month. The nature of this turbidity is unknown. Checker R. S. M. preferred not to boil the reagent since it avoided turbidity. Checker A. K. preferred ceric hydrate as made by the Lindsay Chemical Co. (West Chicago, Ill.) instead of ceric ammonium sulfate because of its greater purity. He dissolved 4.3 g. ceric hydrate in 45 ml. conc. H_2SO_4, brought it to a boil and let it simmer for 30 minutes. After cooling it was added to 500 ml. of water, cooled again, and made to 1000 ml.

Standard iodine solutions.

Stock solution A. 100 μg. iodine per milliliter. Dissolve 118.1 mg. of sodium iodide dried in desiccator and dilute to 1000 ml. Store in a dark bottle in refrigerator.

NOTE: Checker A. K. preferred KIO_3 for the standard because it was more stable in solution.

Stock solution B. 0.1 μg. iodine per ml. Dilute 1.0 ml. of stock solution A to 1000 ml. Store in a dark bottle in refrigerator.

Diluted standard iodide solutions. 0.01, 0.02, 0.03, 0.04, 0.05, and

0.06 µg. iodine per milliliter. Dilute 10, 20, 30, 40, 50, and 60 ml. of stock solution B each to 100 ml. Store in dark bottles.

NOTE: No measurable decrease of the content of iodine in the diluted standard solutions is found within 1 month. It seems, advisable, however, to renew the diluted standard solutions every month.

Procedure

PRECIPITATION AND WASHING

Of each serum two 1-ml. samples (or 0.5 ml. if PBI exceeds 15 µg. per 100 ml.) are pipetted into incineration tubes. To each serum sample is added 7 ml. of water, 1 ml. of zinc sulfate solution, and 1 ml. of 0.5 N sodium hydroxide in the order given. Using the glass rod, mix thoroughly as the NaOH is added. The precipitate that forms is stirred well with this footed glass rod (a separate rod for each tube), allowed to remain at room temperature for 10 or more minutes, and then centrifuged for 10 minutes at about 1000 g (e.g., 2000 RPM with a radius of 15 cm. to the center of the tube). The supernatant liquid is decanted and the precipitate is washed by centrifugation with three successive portions of 10 ml. of iodine-free water. With each portion of water the mixture is thoroughly stirred with the footed rod until the precipitate is broken up into a uniform fine suspension.

NOTE: The precipitated protein, centrifuged down, contains about 27% of the inorganic iodide which was present in the serum. After the first washing this figure is reduced to about 6% and after the second washing it is 2%. After the third washing only 1% of the inorganic iodide is still occluded in the protein. If the precipitate is not broken up into a complete suspension in the water before the centrifuging, the effect of the washing procedure is diminished, and up to 5% of the inorganic iodide may remain in the protein after the third washing. Checker R. S. M. believes that one washing is sufficient if thorough mixing of the reaction mixture is performed while the proteins are being precipitated.

The inorganic iodide can be removed by dialysis of serum [Riggs et al. (3) and Vilkki (7)]. The author has found that approximately 99% of the inorganic iodide is removed after dialysis for 12 hours against running water (1 ml. of serum is placed in an 8 × 100 mm. dialyzer tubing, American Instrument Co., Cat. No. 5–8995). The PBI values obtained are the same if either the technique of dialyzing or the technique of precipitating the proteins is used. If the dialyzing procedure is adopted tap water should not be used for the dialysis because it often contains iodine (0.5 µg. per 100 ml. of water in the submitter's laboratory).

Slade (10) introduced the use of anion exchange resin in order to remove the inorganic iodide. The author has confirmed Slade's report that more than 99% of the iodide is absorbed by the resin. However, the PBI values are a little higher when the resin is used, compared with the values obtained with the precipitation procedure. The author subjected 40 different sera to both methods and found that the resin technique gave as an average 0.8 μg. PBI more per 100 ml. of serum than the precipitation technique.

If the technique of dialysis or the use of resin is chosen, the iodide-free serum should receive 1 ml. of zinc sulfate solution together with the potassium hydroxide. The presence of zinc compounds promotes the incineration and the zinc carbonate precipitate present when the ash is extracted clarifies the extract.

DIGESTION AND INCINERATION

One milliliter of 2.0 N potassium hydroxide is added to each precipitate. The contents of the tubes are dried overnight at 115°C. Then the rack with the tubes is placed in the cold muffle oven. The temperature is raised during about half an hour to 600°C and kept at 600°C ± 10°C for 1 hour. During the period of heating up to 600°C it is important that the door of the oven fits closely and that the ventilation hole is closed so that circulation of air in the furnace is minimized. If the oven is located in a hood, the draft should be shut off. Five, twenty, and forty minutes *after the temperature has reached 600°C* the door is opened for about 15 seconds in order to renew the air in the oven. The incineration is usually completed within 1 hour at 600°C.

NOTE: During the period of heating up from 400° to 600°C iodine-containing smoke is formed. If the smoke escapes from the incineration vessel, iodine is lost. After the temperature has been established at 600°C, however, no loss of iodine occurs. In order to prevent losses of iodine circulation of the air in the oven and tubes is minimized until the temperature reaches 600°C. For this purpose the incineration is carried out in tubes 15 × 125 mm. rather than in vessels of larger diameter. For the same reason the oven is tightly closed during the period of heating up to 600°C. Potassium hydroxide was selected as the alkali for ashing because iodine showed less tendency to escape from it than from sodium hydroxide or sodium carbonate. In the present procedure the loss of iodine is less than 2%, and it has not been found necessary to correct for it in the calculations of the results. When the described technique is used the loss of iodine is still less than 2% even when the incineration is carried out at 650°C. Cross contamination of iodine from one tube to the neighbor tube during the incineration was looked for, but could not be detected.

The combustion of the organic matter of 1 ml. of serum requires approximately 70–90 mg. of oxygen, the amount which is present in 1 l. of air measured at 600°C. When several samples of serum are incinerated at the same time the amount of oxygen in the oven is inadequate to complete the combustion. The most convenient way to supply the combustion with the necessary amount of oxygen is to open the door of the oven for 10–15 seconds at about 15 minutes interval after the temperature has been established at 600°C. No loss of iodine is caused by renewing the air at this step of the incineration.

Preparation of Solution of Ash

Ten milliliters of water are added to the ash in each tube. Using a footed glass rod ash attached to the wall is loosened and the entire contents of the tube are thoroughly stirred. The glass rod is removed without washing it. The tubes are centrifuged for 10 minutes.

NOTE: The procedure of preparing an alkaline extract of the ash has some advantages over dissolving the ash in excess of acid: (1) Any unburned carbon particles remaining from the incineration are centrifuged down with the insoluble part of the ash giving a clear supernatant. This is not the case when the ash is dissolved in excess of acid. (2) No loss of iodine occurs when the alkaline extract is prepared, and in the subsequent acidifying the loss is negligible. When the dry ash is dissolved in excess acid up to 10% of the iodine may be lost during the stormy evolution of CO_2. No iodine is lost by adsorption to the insoluble part of the ash during the preparation of alkaline extract.

NOTE (*important*): If the conditions for the incineration or extraction of the ash differ from those given in this method, the analyst is earnestly advised to measure the loss of iodine during the procedure. This is easily done by measuring the loss of I^{131} added to the mixture of proteins and alkali before the step of drying [e.g., see Foss et al. (8)].

Determination of Iodine by Decolorization of Cerate

Of each clear supernatant solution 2 portions of 4 ml. are transferred to photometer cuvettes. To each cuvette are added 0.5 ml. of 0.1 N arsenite solution and 1 ml. of sulfuric-hydrochloric acid reagent; the acid is added gradually to avoid excessive foaming from evolved CO_2. The cuvettes are then placed in a 37°C water bath for at least 10 minutes to allow them all to reach the bath temperature. In sequence, at 30-second intervals, the cuvettes each receive 1 ml. of ceric sulfate solution. After thorough stirring (individual footed rods), each cuvette is replaced in the bath before the next one receives the ceric sulfate. Twenty minutes after each tube received ceric sulfate the tube is removed from the water bath, wiped dry and the optical density is read at a wavelength of 420 mμ, so that readings follow each other at 30-second intervals. A cuvette filled with water is used to set the photometer at zero optical density. For each reading the time interval between addition of ceric sulfate and reading, viz., 20 minutes plus the time required to take the cuvettes from the bath and take the reading, is kept as constant as possible, e.g., at 20 minutes and 20 seconds.

NOTE: For optimal accuracy of the photometric measurement it is desirable to have the readings of the optical densities within the limits where they have the best accuracy, viz., between 0.70 and 0.15, and to have the conditions of catalysis so chosen that the maximum amount of iodine for which the method is intended will cause a fall of the optical density from about 0.70 to about 0.15. As the rate of the reaction is influenced by impurities, such as iodine and heavy metals, present in the reagents, and because different types of photometers give different optical density for the same ceric solution, the time of reaction given in this procedure, 20 minutes, will not always be the most favorable. It is advisable to prepare standard curves as outlined below with the exception that the optical densities are read after 10, 15, 20, 25, and 30 minutes, and from the results to decide which reaction time is preferable.

PREPARATION OF STANDARD CURVE

Immediately before or after the preparation of each series of ash solutions, a standard curve is prepared. Into seven cuvettes are measured 1-ml. portions, respectively, of water and of standard solutions containing 0.01, 0.02, 0.03, 0.04, 0.05, and 0.06 μg. of iodine per milliliter. To each cuvette are added 0.4 ml. of 2 N potassium hydroxide, 0.5 ml. of 0.1 N sodium arsenite solution and 2.6 ml. of water, making a total volume of 4.5 ml. To each cuvette is then added 1 ml. of sulfuric-hydrochloric acid reagent, the solution is warmed to 37°C, 1 ml. of ceric sulfate solution is added and the optical densities are read after 20 minutes in series exactly as directed for the ash solution. On semilogarithmic paper the extinctions of the standard are plotted against micrograms of iodine (the optical densities along the logarithmic axis). (Fig. 2).

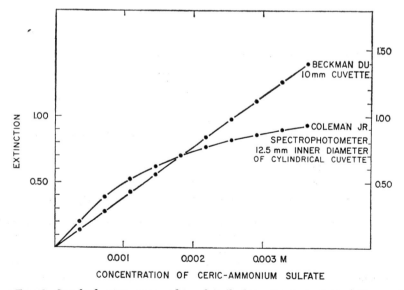

FIG. 2. Standard curve prepared as described in the text. Optical densities measured in a Coleman Junior Photometer at 420 mμ.

NOTE: A standard curve should be prepared simultaneously with each series of analyses because standard curves prepared at different times differ somewhat. The greatest difference found between ten standard curves prepared within one month in the submitter's laboratory corresponded to 0.5 μg. of PBI per 100 ml. of serum. The incineration is omitted when preparing the standard curves, because standard curves are very nearly the same whether the standard solutions are incinerated or not.

PREPARATION OF BLANKS

With each series of sera two reagent blanks, each containing 1 ml. of iodine-free water are prepared. The blanks go through the same procedure as the samples of serum, viz., addition of H_2O, $ZnSO_4$, and NaOH, centrifuging and washing the precipitate three times, addition of KOH, drying, incineration, and determination of the content of iodine in the solution of the ash.

NOTE: In the author's laboratory the blank values usually correspond to 0.25–1.0 μg. of iodine per 100 ml. of serum. If they are higher it is advisable to look for contaminations.

NOTE (*important*): The presence of protein retards greatly the velocity of the reaction between ceric ions and arsenite as catalyzed by iodine. The presence of only 0.002 ml. of serum completely prevents the reaction and even 0.0001 ml. serum or 0.01 ml. saliva interferes markedly. The most rigid precautions must be taken in order to prevent contamination with organic matter after the incineration is completed.

Calculation

Calculation is by the equation:

$$PBI = \frac{250(S\text{-}B)}{V_s}$$

PBI = μg. of PBI per 100 ml. of serum.

S = μg. of iodine found in the solution of ash from 0.4 of the ashed sample.

B = μg. of iodine found in the blank.

V_s = ml. of serum (usually 1 ml.) in the sample ashed.

S and B are obtained by interpolation on a standard curve.

NOTE: Beer's law is obeyed only when ceric sulfate solutions are analyzed in spectrophotometers which, like Beckman DU, provide nearly monochromatic light. When simpler photometers are used the observed optical densities differ considerably from Beer's law and the obtained standard curve is very far from a straight line. With such a photometer the values of S and B in the above equation must be determined by interpolation between observed optical densities on a standard curve prepared from a series of accurately determined points.

Normal Serum PBI

The concentration of protein-bound iodine in a group of 55 healthy volunteers was found to range from 3.5 to 8.0 μg. per 100 ml. of serum (8). The average was 5.56 μg. No significant difference in PBI concentration was found between fasting and nonfasting subjects. The mean of 30 fasting samples was 5.52 μg. per 100 ml. of serum, that of 25 nonfasting was 5.61 μg. per 100 ml. of serum (8).

During pregnancy PBI is somewhat higher. Heinemann et al. (11) found 11.2 μg. per 100 ml. serum as the upper limit in a group of healthy pregnant women.

The serum PBI of newborn infants is distinctly higher than that of adults during the first week of life, ranging from 7 to 14 μg. per 100 ml. The level decreases rapidly in the first 3 months and then slowly reaches the mean level for adults in early childhood.

Precision of the Determination of PBI

Based upon separate analyses of ten replicate samples from each of two serum pools and preparation of ten standard curves, a standard deviation of ± 0.4 μg. PBI per 100 ml. of serum was found when the content of PBI was about 5 μg. per 100 ml. of serum and ± 0.5 μg. PBI when the serum contained about 10 μg. PBI per 100 ml. (8). When the average of duplicate analyses is taken these figures are reduced to approximately ± 0.3 and ± 0.35 μg. PBI per 100 ml. of serum.

Recovery of Thyroxine Added to Serum

Thyroxine was added to different pools of serum in amounts which increased the concentration of PBI with from 4 to 8 μg. per 100 ml. serum (8). From six to ten samples from each pool were analyzed. The recovery was found to range from 98 to 103%.

Discussion

Kinetics of the iodide-accelerated reaction between Ce[iv] and As[iii]. As shown by Acland (12), in the absence of interfering substances the time course of the reaction runs according to the first-order reaction equation:

$$K_1 + K_2 I = \frac{1}{t} \ln \frac{C_o}{C_t} = \frac{1}{t} (\ln C_o - \ln C_t) \qquad (1)$$

K_1 is the reaction constant for the reduction rate of Ce^{IV} by As^{III} in the absence of added iodide, I is the concentration of added iodide, K_2 is the catalytic coefficient for the added iodide, t is the duration of the reaction, C_o is the initial concentration of Ce^{IV}, C_t is the concentration of Ce_{IV} after the reaction has run for t minutes following addition of iodide. If t is taken as a constant (e.g., at 20 minutes as in the present procedure) the equation, for a given initial C_o, reduces to a linear relation between I and either in C_t or $\log_{10} C_t$.

$$I = a - b \log C_t \tag{2}$$

When the photometer used gives a linear relation of Ce^{IV} to optical density, D, the log of D has a corresponding linear relation to added iodide.

$$I = c - d \log D \tag{3}$$

where $a,b,c,$ and d are composite constants.

The linearity of Eq. (3) can be destroyed by failure of the photom-

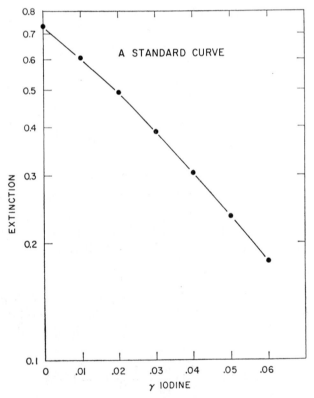

FIG. 3. Relation between the concentration of ceric ammonium sulfate dissolved in 1.5 N sulfuric acid and optical density measured at 420 mμ in a Beckman Spectraphotometer DU and in a Coleman Junior Spectrophotometer.

eter used to give results according to Beer's law, or by impurities in the reagents which affect the course of the reaction. Acland (12) found that the zinc he used appeared to retard the reaction when the amounts of iodide added were in the upper range used, so that the curve of optical density versus iodide was somewhat convex. In Fig. 3 the curve is slightly concave from the failure of the Coleman Jr. photometer to give a linear relation between ceric concentration and optical density.

When conditions validate linear Eq. (3) it can be used to calculate iodine values from log D, the constants c and d being determined by analysis of standard iodide solutions. But calculation based on assumption of linearity between optical density itself and iodide concentration can give only approximate accuracy over a limited range. If, however, an empirical standard curve is used to interpolate iodide concentration it does not matter whether the curve relates iodide concentration to log D, as done by Barker (5) and Acland (12), or to D, as done by Skanse and Hedenskog (13) and Kingsley (14); an accurate empirical standard curve of any of these relations will serve for accurate graphic calculation of iodine values whether the curve is linear or bent.

REFERENCES

1. Sandell, E. B., and Kolthoff, I. M., Microdetermination of iodine by a catalytic method. *Microchim. Acta* 1, 9–25 (1937).
2. Trevorrow, V., Studies on the nature of the iodine in blood. *J. Biol. Chem.* 127, 737–750 (1939).
3. Riggs, D. S., Lavietes, P. H., and Man, E. B., Investigations on the nature of blood iodine. *J. Biol. Chem.* 143, 363–372 (1942).
4. Salter, W. T., and McKay, E. A. Iodine in blood and thyroid of man and small animals. *Endocrinology* 35, 380–390 (1944).
5. Barker, S. B., and Humphrey, M. J., Clinical determination of protein-bound iodine in plasma. *J. Clin. Endocrinol.* 10, 1136–1141 (1950).
6. Grossmann, A., and Grossmann, G. F., Protein-bound iodine by alkaline incineration and a method for producing a stable cerate color. *J. Clin. Endocrinol.* 15, 354–361 (1955).
7. Vilkki, P., A modification of the alkaline combustion method for the determination of protein-bound iodine in serum. *Scand. J. Clin. Lab. Invest.* 10, 272–277 (1958).
8. Foss, O. P., Hankes, L. V., and Van Slyke, D. D., A study of the alkaline ashing method for determination of protein-bound iodine in serum. *Clin. Chim. Acta*, 5. 301–326 (1960).
9. Somogyi, M., A method for the preparation of blood filtrates for the determination of sugar. *J. Biol. Chem.* 86, 655–663 (1930).
10. Slade, C. I., Application of a simple ion exchange resin technique to determination of serum or plasma protein-bound I^{127} or protein-bound I^{131}. *J. Clin. Endocrinol.* 16, 1122–1125 (1956).

11. Heinemann, M., Johnson, C. E., and Man, E. B., Serum precipitable iodine concentrations during pregnancy. *J. Clin. Invest.* **27**, 91–97 (1948).
12. Acland, J. D., The estimation of protein-bound iodine by alkaline incineration. *Biochem. J.* **66**, 177–188 (1957).
13. Skanse, B., and Hedenskog, I., The determination of serum protein-bound iodine by alkaline incineration. Values in normal subjects. *Scand. J. Clin. Lab. Invest.* **7**, 291–297 (1955).
14. Kingsley, G. R., and Schaffert, R. R., Protein-bound iodine in serum. *Std. Methods Clin. Chem.* **2**, 147–164 (1958).

SERUM IRON AND SERUM IRON-BINDING CAPACITY*

Submitted by: THOMAS J. GIOVANNIELLO, Laboratory Service, Boston Veterans
Administration Hospital, Boston, Massachusetts

THEODORE PETERS, JR., Mary Imogene Bassett Hospital, Cooperstown, New York

Checked by: EDWARD WAGMAN, Veterans Administration Hospital, West Haven,
Connecticut

I. SERUM IRON

Introduction

Methods for determination of serum iron generally employ as a
first step some technique for the release of iron from its combination
with the iron-binding globulin. Use of dilute (0.2–0.35 N) HCl, introduced originally by Barkan (1), is a simple procedure, but has been
found to give incomplete (75–80%) extraction of iron from serum
when applied to specimens which had been stored in the frozen state
(2). Since in most laboratories the determination of serum iron is not
performed daily, it is of importance that the method used be equally
valid for fresh or stored specimens.

By addition of a reducing agent together with the HCl, the method
described by Barkan and Walker (3) and Schales (4) has been found
to give complete recovery of iron from fresh or stored specimens. The
method presented below, originally described by Peters, Giovanniello,
Apt, and Ross (2), uses thioglycolic acid as the reducing agent. Proteins are then precipitated by trichloroacetic acid, the pH is adjusted
to 4.3 by addition of sodium acetate, and iron is determined as the
iron-bathophenanthroline complex, which gives more than twice as
much color as the complex with 1,10-phenanthroline or a-a'-dipyridyl.
The bathophenanthroline is added in isopropyl alcohol due to its in-
solubility in water. Further addition of reducing agent is unnecessary
since the iron is already in the ferrous state.

* Based on the method of Peters et al. (2).

Reagents

NOTE: Use of iron-free glassware (syringes, pipets, test tubes, stirring rods) is necessary throughout. Glassware is made iron-free by soaking in 6 N HNO₃ then rinsing thoroughly in iron-free water. De-ionized water is usually excellent for such use. Simple distilled water should be further distilled in all-glass apparatus or passed through a de-ionizer if any color is obtained with batho-phenanthroline and reducing agent. Hypodermic needles are not specially treated. Rubber stoppers are avoided. Use glass stoppers or paraffin film. Iron-free water should be used in making reagents.

1. Trichloroacetic acid solution, C.P., 30%.

2. Thioglycolic acid (2-mercaptoethanoic acid, HSCH₂COOH) C.P., 80% solution in water. Store in a glass-stoppered dropping bottle.

NOTE: For best results, the trichloroacetic acid and thioglycolic acid reagents should be distilled to make them iron-free. However, many batches of C.P. material have been found to give a sufficiently low blank without further purification. These two reagents are easily contaminated with use and should be suspected if high blanks are encountered.

3. Approximately 0.20 N hydrochloric acid. Dilute 17 ml. of concentrated HCl (A.R.) to 1 l. with water.

4. Sodium acetate (NaAc) solution, approximately 4.3 N (35%). Solutions saturated with sodium acetate (A.R.) at room temperature have been found satisfactory, and can be conveniently prepared simply by adding water to sodium acetate in its original bottle. The temperature of saturation has been found not to be critical between 3°C. and 40°C.

5. Bathophenanthroline (4,7-diphenyl-1,10-phenanthroline), 0.020% solution in isopropyl alcohol, C.P.

NOTE: Bathophenanthroline is available from the G. Frederick Smith Chemical Company, Columbus, Ohio. It dissolves slowly and should be allowed to stand, preferably with agitation, at least overnight.

Standardization

1. Stock solution. Weigh about 1 g. of pure iron wire to within 1 mg. and dissolve in dilute HCl. Dilute to 1 l. with water.

2. Working standards. Dilute 1 ml. of the stock standard to 500 ml. with water. This solution contains about 200 μg. of Fe per 100 ml. A series of working standards can then be made by further dilution of this solution to check linearity of the standard curve. Use 2.0 ml. to replace the serum in the procedure outlined below. The working standards must be prepared just before use. The standard factor need be determined only occasionally, since it remains constant for a par-

ticular instrument unless the characteristics of the instrument change. Analysis of standard iron solutions incorporates the various dilutions inherent in the method, so that these need not be considered when the standard factor is determined in this manner. One small correction, however, should be applied. When the serum proteins are precipitated by trichloroacetic acid, the resulting 6 ml. suspension contains about 1.5% solids, so that the total liquid volume is slightly less than 6.0 ml., and the 4.0-ml. aliquot removed is slightly more than two-thirds of the total liquid. The value of this correction factor was found to be 1.015 in studies of recovery of Fe^{59} (2). Since it applies to the determination of serum but not of standard solutions, the standard factor becomes:

$$\frac{\mu g.\ Fe\ per\ 100\ ml.\ in\ solution\ analyzed}{net\ optical\ density \times 1.015} = standard\ factor.$$

For example, analysis of a solution containing 200 μg. Fe per 100 ml. gave a net optical density of 0.213. The standard factor for that particular instrument is:

$$\frac{200}{0.213 \times 1.015} = 925.$$

The optical density at 535 mμ of a solution containing 1 μg. of Fe per ml. is 0.401 on the Beckman DU spectrophotometer in a 1.000 cm. cuvette, corresponding to a molar absorbancy index of 22,400. On the Coleman Junior spectrophotometer in 19-mm. round cuvettes the corresponding optical density is 0.518.

Specimens

Blood is drawn with an iron-free syringe and handled in iron-free glassware only. Serum or citrated, oxalated, or heparinized plasma is satisfactory. Note, however, that citrated or oxalated plasma is not suitable for determination of serum iron-binding capacity (Part II). Specimens which show visible hemolysis will give slightly high results. Serum or plasma may be analyzed immediately or stored in the frozen state.

Procedure

Pipet 2.0 ml. of serum or plasma into a 15-ml. test tube. Prepare a blank using 2.0 ml. of water. Add 3.0 ml. of 0.20 N HCl and 1 drop of thioglycolic acid. Mix by swirling and let stand 20–40 minutes at room temperature.

Add 1.0 ml. of 30% trichloroacetic acid. Mix with a stirring rod and let stand 15–30 minutes at room temperature, covered with paraffin film. Centrifuge for about 15 minutes at about 1500 × g.

Pipet 4.0 ml. of the supernatant into a cuvet. Use of pipet controller will permit removal of this amount of supernatant without disturbing the precipitate. Add 0.5 ml. of 35% sodium acetate and 2.0 ml. of the alcoholic bathophenanthroline solution. Mix by swirling.

NOTE: A modification reported by Trinder (5) may serve to increase the sensitivity of the method further. He obtained the water-soluble sulfonated derivative of bathophenanthroline by boiling 100 mg. of bathophenanthroline with 0.5 ml. of chlorosulfonic acid for 30 seconds, then cooling, adding 10 ml. of water cautiously, and heating to 100°C. until the precipitate dissolved. The solution was diluted to 100 ml. and 0.2 ml. used in the iron determination. If applied to the method described here, in order to have a final volume of 5.0 ml. for use in standard spectrophotometers, it would be preferable to dilute the sulfonated bathophenanthroline to 250 ml. instead of 100 ml., and add 0.5 ml. This would give a final volume of 5.0 instead of 6.5 ml., with a concomitant 30% increase in sensitivity.

Read the optical density against water after 5 minutes or more at 535 mμ or with a green filter. The color is stable for at least 24 hours.

Calculations

Subtract the optical density of the blank from that of the unknown. Multiply by the previously determined standard factor. Results are obtained as μg. Fe per 100 ml.

Example:

Optical density of blank	0.012
Optical density of unknown	0.129
Standard factor	925

$(0.129 - 0.012) \times 925 = 108 \ \mu$g. Fe per 100 ml. serum.

The standard curve is linear to a serum iron concentration of slightly over 300 μg. per 100 ml. Should a serum be encountered with a higher iron content than this, the analysis should be repeated using 1 ml. of serum and 1 ml. of water.

Discussion

The blank tube, read against water, should not exceed an equivalent serum iron concentration of 10 to 15 μg per 100 ml. (optical density of about 0.012 in 19-mm. round cuvette).

The use of reducing agents as strong as thioglycolic acid has been avoided by some workers because under some conditions it extracted iron from hemoglobin. Tests on the method presented showed that the presence of 30 mg. per 100 ml. of hemoglobin caused an increase in the serum iron value of only 5 μg. per 100 ml., corresponding to extraction of about 5% of the iron from the hemoglobin. This amount of hemolysis is detectable to the unaided eye, hence specimens not visibly pink can be safely analyzed by this method.

The specificity of this procedure for iron is shown by the agreement of the absorption spectra of solutions resulting from determination of iron in serum with those resulting from standard iron solutions (2). No copper was found to be extracted from serum by the conditions used. Cobalt gave no color with bathophenanthroline.

Recovery of Fe^{59} added to serum *in vitro* or *in vivo* was found to be 101.5% ± 0.48% in a large series of fresh and stored specimens (2). That this value was slightly greater than 100% was believed to be due to the effect of precipitation of the proteins as a separate phase as mentioned previously. This effect is corrected for in computing the standard factor. Precision of 24 duplicate determinations was about 1 μg. per 100 ml.

Sera from a series of forty-three normal males gave an average value of 119.7 μg. per 100 ml., with a range of 56–183. This is in the midrange of the values reported by others (6, 7, 8, 9, 10), indicating the validity of the procedure despite objections which have been raised that the use of bathophenanthroline gives high values (4). Females average about 10 μg. per 100 ml. lower than males, and small children are somewhat lower still, ranging from 60–100 μg. per 100 ml. between 6 months and 2 years (10).

Attempts to establish a diurnal variation have not given uniform results (9, 11), although in some individuals the level is apparently higher in the early morning than in the afternoon, due to slowing of hematopoiesis at night.

Values in pathological states have been discussed by Schales (4) and for further information the reader is referred to Wintrobe's text (12) and to the articles by Vahlquist (6) and Laurell (8).

In the published methods for the determination of serum iron many techniques have been used for the initial release of iron from the serum. The bond to the iron-binding globulin can actually be broken simply by lowering the pH to 5, but the released iron precipitates with the proteins when they are removed unless more drastic measures are taken. Usually acid, reducing agents, or heat are used alone or in combination. Strong HCl (1–6 N), used particularly by German and Scandinavian workers (6, 8, 12, 13, 14, 15, 16, 17), has been found to give complete extraction from fresh or stored specimens, but requires large amounts of alkali for neutralization. Procedures using dilute HCl (1, 3, 4, 18) may not extract all of the iron from stored specimens. Somewhat variable results have been reported (2, 19, 20) with procedures using sulfite (19, 21), thioglycolic acid alone (20), or trichloroacetic acid plus heat (5, 22). Hydroxylamine and heat has been used (23), as well as removal of iron without precipitation of the serum proteins with ascorbic acid at pH 6 (24) or by ultrafiltration (6). Recently, tetravalent thorium has been employed to displace iron from its combination with serum proteins at alkaline ph (25). A thorough review of methods appearing before 1940 has been made by Vahlquist (6).

For development of color thiocyanate, a-a'-dipyridyl, terpyridine, o-phenanthroline, and bathophenanthroline have been employed. The latter is the most sensitive, but suffers from its low solubility. The water-soluble sulfonated derivative (5) may prove the optimal reagent. Potentially even more sensitive methods have utilized the catalysis by iron of oxidation of cysteine (26) or o-tolidine (27), but these are difficult in application.

II. SERUM IRON-BINDING CAPACITY

Introduction

The ability of serum to combine with iron is known as Iron-Binding Capacity (IBC). Upon addition of small increments of iron a maximum is reached where no more iron will combine with the iron-binding protein (also referred to as transferrin or siderophilin). The determination of serum iron content is a measurement of iron already bound to protein, while the determination of the iron-binding capacity is a measurement of the iron that can be bound by the same protein. The amount which can be bound in excess of the iron content is sometimes referred to as latent or unsaturated iron-binding capacity (UIBC). The sum of the iron content and the UIBC is the total iron-binding capacity (TIBC).

Published methods using optical procedures have proven unsatisfactory with sera which are not optically clear. Changes in physical characteristics of sera, such as color or density, may result from vary-ing dietary intake or storage. Such sera cannot be accurately analyzed by optical methods since their spectral absorptions are altered.

Attempts were made (28) to obtain saturated serum by adding iron as the ferrous ion and removing the unbound excess by cation ex-change resins. Tests with a number of resins under a variety of con-ditions were unsuccessful. Either the resin failed to remove all of the excess iron or it removed some of the iron bound by the serum protein. However, an anion exchange resin, Amberlite IRA 410, was found to remove excess iron when the iron was added as ferric ammonium citrate and not to remove any of the protein-bound iron.

Ion-exchange resins such as the one used in this method are syn-thetic polymers which, though insoluble, simulate or react much as acids, bases, or salts. These resins might be considered as macro-molecules or polyelectrolytes. Ion-exchange resins are able to ex-change their cations (or anions) for other cations (or anions) in aqueous and nonaqueous media. An example would be the exchange of sodium ions for some other cation, such as calcium, which it was desired to remove from a solution. Amberlite IRA 410 is a strongly basic anion exchanger which has an active quaternary amino group and which is beadlike in shape. The action of the anion-exchange resin in this method is apparently to bind one of the carboxyl groups of citrate in exchange for chloride, and the iron remains attached to the other carboxyl groups of the citrate.

NOTE: Data describing resins have been taken from technical bulletins pub-lished by the Rohm and Haas Company, Washington Square, Philadelphia 5, Pennsylvania.

In the method described in this article (28), 1 ml. of serum is satu-rated by adding an excess of ferric ammonium citrate. The iron which is not bound by the serum proteins is then removed by addition of a small amount of IRA 410 buffered at pH 7.5 in the chloride cycle. After dilution and centrifugation an aliquot of the supernatant is analyzed for iron content giving a measure of the total iron-binding capacity of the serum.

Reagents

1. Ferric ammonium citrate, 0.050 mg. Fe/ml. Since commercial preparations appear to contain some iron as the hydroxide this reagent is prepared as follows: In a 50-ml. centrifuge tube, 60 mg. $FeCl_3 \cdot 6H_2O$

dissolved in 20 ml. of distilled water is precipitated with 10–20 drops of 6 N ammonium hydroxide. The tube is centrifuged and the precipitate is dissolved with a few crystals of citric acid with heat. Then the pH is adjusted to 6.5–7 with 6 N and 1 N ammonium hydroxide using bromthymol blue indicator. The iron solution is transferred quantitatively to a 250-ml. volumetric flask and diluted to volume. The solution is kept under refrigeration and appears to be stable, as indicated by a light green color, for at least 6 months.

2. *Buffer, pH 7.5, containing 0.11 M NaCl and 0.044 M barbital (diethylbarbituric acid).* Due to the low solubility of barbital at this pH, it is helpful to approach the pH from the alkaline side. Use 6.4 g. of NaCl, 6.0 g. of diethylbarbituric acid and 2.3 g. of its sodium salt per liter. Store at room temperature.

3. *Concentrated hydrochloric acid, (A.R.), in glass-stoppered dropping bottle.*

4. *Amberlite IRA 410 resin (Rohm and Haas, Philadelphia 5, Pennsylvania).* The resin is suspended in approximately 3 N HCl overnight to wash it free of iron and to saturate it with chloride; then it is washed well with distilled water. It is brought to pH 7.5 (phenol red indicator) by suspension in the above buffer, with the addition of a little NaOH if necessary. It is then dried at 95°C.

5. *Reagents described for determination of serum iron, in the first part of this chapter, except that the 0.2 N HCl is not required.* The chapter on serum iron determination describes the method for preparing iron-free apparatus.

Specimens

Specimens obtained for the determination of iron-binding capacity must be free of oxalate or citrate. Anticoagulants of this nature will bind with the resin releasing calcium ions and cause the specimen to clot. For this reason only serum or heparinized plasma should be used. This method is applicable to fresh or stored sera and to lipemic or icteric specimens. Ordinary refrigeration or freezing is suitable for storing. Use glass stoppers or paraffin film for covering specimens during storage.

Procedure

Pipet a 1.0-ml. sample of serum or heparinized (not citrated or oxalated) plasma into a centrifuge tube. Add 0.10 ml. of the ferric ammonium citrate solution (equals 5 μg. of iron). Let stand about 10 minutes at room temperature.

Add 0.4 ml. of the dry resin and stir occasionally for 5 to 10 minutes.

NOTE: The resin is measured by volume for convenience. A glass spoon holding 0.4 ml. can be prepared locally, or a calibrated 12-ml. centrifuge tube can be used.

Then add 5.0 ml. of buffer and again stir occasionally for 5 to 10 minutes. Centrifuge briefly and remove 5.0 ml. of the supernatant using a manual pipet control to avoid disturbing the resin.

Place the 5 ml. of the supernatant in a 1.5 × 12.5 cm. tube and analyze for iron content as follows.

Add 1 drop of concentrated HCl and 1 drop of thioglycolic acid. Mix by swirling and let stand 30 minutes at room temperature.

Add 1.0 ml. 30% trichloroacetic acid. Proceed as described from this point for serum iron determination in Part I of this chapter.

Calculations

Reference is made to the discussion in Part I on Standardization and Calculations, with the following additions:

Only 1.0 of serum is used and a five-sixths aliquot of original sample is taken. The amount of solid phase present upon precipitation of the protein is less, so that the correction factor is 1.006 instead of 1.015. Therefore, if analysis of a solution containing 200 μg. Fe per 100 ml. gives a net optical density of 0.213, the standard factor for TIBC is:

$$\frac{200 \times 2 \times 6}{0.213 \times 1.006 \times 5} = 2240.$$

Example of calculation:

Optical density of blank   = 0.015
Optical density of unknown = 0.202
Standard factor   = 2240

$(0.202 - 0.015) \times 2{,}240 = 419$ μg. Fe per 100 ml. (TIBC)

NOTE: The value obtained for serum iron subtracted from TIBC gives unsaturated or latent iron-binding capacity (UIBC) if desired.

Discussion

Any amount of resin between 0.4 and 1.0 ml. will remove the excess iron. The use of too much resin to remove the excess iron will cut down on the yield of supernatant, making it difficult to obtain the proper aliquot needed for analysis.

It is suggested that a blank be run with each analysis. This should be made with the use of 1 ml. of water in place of serum. Blank values due to resin treatment should not amount to more than 10–15 μg. per 100 ml.

The precision of the method is illustrated by the results of eleven analyses made on the same serum specimen (28). Range of values was 282 to 315 μg. per 100 ml., with a standard deviation of 10 μg. per 100 ml.

A series of normal and pathologic sera was analyzed by the optical method of Rath and Finch (29) and by the resin method. Serum iron content was determined by the method described in Part I. The agreement between the resin and optical methods was about ±30 μg. per 100 ml. Results with the resin method tended to be slightly higher, but the average difference of 23 μg. per 100 ml. is of doubtful significance.

A series of thirty-four normal sera analyzed by the resin method gave an average normal value of 333 μg. per 100 ml. This is in good agreement with the results obtained by others with other methods (8, 14, 29, 30, 31, 32). Two-thirds of normal sera showed TIBC's in the range of 277–379 μg. per 100 ml., but the extreme values ranged from 248–422 μg. per 100 ml.

Results on pathologic specimens were in agreement with findings of others (8, 14, 29, 30, 31, 32). TIBC was found to be elevated in iron-deficiency anemia and reduced in hemochromatosis, nephrotic syndrome, and malignancy.

Diurnal-nocturnal variations have been found to be insignificant (11).

Iron labeled with a radioactive tracer has been used (33, 34) to saturate serum for the determination of iron-binding capacity. The procedure of Feinstein et al. (33) was investigated, and the results reported were consistently 35–115 μg. per 100 ml. higher than optical methods (28). In Tinguely's method (34) ammonium sulfate is used to separate the serum proteins followed by the addition of ethanol. The use of ethanol precipitates the ammonium sulfate colloidal suspension which facilitates separation by centrifugation. The procedure of Tinguely was not investigated, but, because the use of filter paper is eliminated, it is assumed that iron-binding values with this method would be lower than the elevated values obtained using Feinstein's procedure.

NOTE: Although methods using radioisotopes are sometimes simpler than chemical methods, the latter are the procedures of choice. The determination of iron-binding capacity as described is designed for use in clinical laboratories.

REFERENCES

1. Barkan, G., Eisenstudien. Die Verteilung des leicht abspaltbaren Eisens zwischen Blutkorperchen und Plasma und sein Verhalten unter experimentallen Bedingungen. Z. Physiol. Chem. 171, 194–221 (1927).
2. Peters, T., Giovanniello, T., Apt, L., and Ross, J. R., A simple improved method for the determination of serum iron. J. Lab. Clin. Med. 48, 280–288 (1956).
3. Barkan, G., and Walker, B., Determination of serum iron and pseudohemoglobin iron with o-phenanthroline. J. Biol. Chem. 135, 37–42 (1940).
4. Schales, O., Serum iron. In "Standard Methods of Clinical Chemistry," Vol. II, (D. Seligson, ed.) pp. 69–78, Academic Press, New York, 1958.
5. Trinder, P., The improved determination of iron in serum. J. Clin. Pathol. 9, 170–172 (1956).
6. Vahlquist, B., Das Serumeisen. Acta Paediat. (Suppl. 5) 28, 1–374 (1941).
7. Cartwright, G., Huguley, C., Ashenbrucker, H., Fay, J., and Wintrobe, M., Studies on free erythrocyte protoporphyrin, plasma iron and plasma copper in normal and anemic subjects. Blood 3, 501–525 (1948).
8. Laurell, C., Studies on the transportation and metabolism of iron in the body. Acta Physiol. Scand. (Suppl. 46) 14, 1–129 (1947).
9. Hamilton, L., Gubler, C., Cartwright, G., and Wintrobe, M., Diurnal variation in the plasma iron level of man. Proc. Soc. Exptl. Biol. Med. 75, 65–98 (1950).
10. Schapira, G., Schapira, F., and Dreyfuss, J., Serum iron, sideropenia experimental high serum iron and siderophilin. Semaine Hop., Pathol. et biol., Fiche tech. biol. 32, (1956).
11. Stengle, J., and Schade, A., Diurnal-nocturnal variations of certain blood constituents in normal human subjects: plasma iron, siderophilin, bilirubin, copper, total serum protein and albumin, haemoglobin and hematocrit. Brit. J. Haematol. 3, 117–124 (1957).
12. Wintrobe, M., "Clinical Hematology" (4th edition), Lea & Febiger, Philadelphia, Pennsylvania (1956).
13. Heilmeyer, L., and Plotner, K., Das serumeisen und die Eisenmangel-Krankheit. Jena 92 (1937).
14. Hagberg, B., The iron binding capacity of serum in infants and children. Acta paediat. Scand. (Suppl.) 93, 1–80 (1953).
15. Matsubara, T., A new method of serum iron determination and commentary on various methods of determination. Kumamoto Med. J. 8, 81–92 (1955).

16. Burch, H., Lowry, O., Bessey, O., and Berson, B., The determination of iron in small volumes of blood serum. *J. Biol. Chem.* **174,** 791–802 (1948).
17. Josephs, H., Determination of iron in small amounts of serum and whole blood with the use of thiocyanate. *J. Lab. Clin. Med.* **44,** 63–74 (1954).
18. Kingsley, G. R., and Getchell, G., Serum iron determination. *Clin. Chem.* **2,** 175–183 (1956).
19. Ramsay, W., The determination of iron in blood plasma or serum. *Biochem. J.* **53,** 227–231 (1953).
20. Peterson, R. E., Improved spectrophotometric procedure for determination of serum iron. *Anal. Chem.* **25,** 1337–1339 (1953).
21. Ramsay, W., An improved technique for the determination of plasma iron. *Biochem. J.* **57,** xvii (1954).
22. Kitzes, G., Elvehjem, C., and Schuette, H., The determination of blood plasma iron. *J. Biol. Chem.* **155,** 653–660 (1944).
23. Koshikawa, H., and Konno, K., Measurement of serum iron. *Sogo Igaku* **10,** 804–805 (1953).
24. Schade, A., Oyama, J., Reinhart, R., and Miller, J., Bound iron and unsaturated iron-binding capacity of serum: rapid and reliable determination. *Proc. Soc. Exptl. Biol. Med.* **87,** 443–448 (1954).
25. Ressler, N., and Zak, B., Simultaneous determination of serum copper and iron by means of the addition of thorium ion. *Am. J. Clin. Pathol.* **28,** 549–556 (1957).
26. Warburg, O., and Krebs, H., Uber locker gebundenes kupfer und eisen im blutserum. *Biochem. Z.* **190,** 143–149 (1927).
27. Budtz-Olsen, O., Micro-estimation of plasma iron with orthotolidine. *J. Clin. Pathol.* **4,** 92–98 (1951).
28. Peters, T., Giovanniello, T., Apt, L. and Ross, J. R., A new method for the determination of serum iron-binding capacity. *J. Lab. Clin. Med.* **48,** 274–279 (1956).
29. Rath, C., and Finch, C., Measurement of iron-binding capacity of serum in man. *J. Clin. Invest.* **28,** 79–85 (1949).
30. Cartwright, G. E., Black, P., and Wintrobe, M. M., Studies on the iron-binding capacity of serum. *J. Clin. Invest.* **28,** 86–98 (1949).
31. Gitlow, S., and Beyers, M., Metabolism of iron. I. Intravenous iron tolerance tests in normal subjects and patients with hemochromatosis. *J. Lab. Clin. Med.* **39,** 337–346 (1952).
32. Houston, J., Haemochromatosis, *Brit. Med. Bull.* **13,** 129–131 (1957).
33. Feinstein, A., Bethard, W., and McCarthy, J., A new method, using radio-iron, for determining the iron-binding capacity of human serum. *J. Lab. Clin. Med.* **42,** 907–914 (1953).
34. Tinguely, R., and Loeffler, K., Method for determining iron-binding capacity of serum. *Proc. Soc. Exptl. Biol. Med.* **92,** 241–242 (1956).

GENERAL REFERENCE

Varley, Harold, "Practical Clinical Biochemistry." Wiley (Interscience), New York (1954).

DETERMINATION OF URINARY NEUTRAL 17-KETOSTEROIDS*

Submitted by: RALPH E. PETERSON, Cornell Medical College, New York, New York

Checked by: ALFRED ZETTNER, Yale University, School of Medicine, New Haven, Connecticut

Introduction

The urinary 17-ketosteroids comprise a group of compounds which have in common a ketone group at position 17 of the steroid nucleus. Some have weakly androgenic properties (androsterone and dehydro-isoandrosterone) whereas others, such as etiocholanolone, are not androgenic. Certain of the urinary estrogens are also 17-ketosteroids but have a phenolic ring and thus can be removed from the urinary extracts with an alkaline wash. The colorimetric methods most widely used for the determination of 17-ketosteroids are modifications of the basic reaction discovered by von Bitto (1) and adapted to steroid ketones by Zimmermann (2). The reaction depends upon the development of a red-purple color with an absorption maximum at 520 mμ when the steroid containing a CH_2CO group is reacted with m-dinitrobenzene in alkali.

Principle

The urinary 17-ketosteroids are excreted as water-soluble sulfate and glucuronide conjugates and these are hydrolyzed by boiling in strong acid. The free steroids are extracted from the hydrolyzed urine with an organic solvent, and the extract treated with m-dinitrobenzene in the presence of alkali. The red color obtained with the 17-ketosteroids does not appear to be much affected by substitutions on other parts of the steroid ring. The 3, 11, and 20 ketosteroids give some color with alkaline m-dinitrobenzene but very much less than the 17-ketosteroids, and their absorption maximum is not at 520 mμ.

* Based on the method of Callow et al. (20) and other modifications.

151

Reagents

1. Petroleum ether (35–65°C. B.P.)–Benzene (1 : 1). Both of these reagent grade solvents are purified separately by passage through a 7 $\times$ 130 cm. column of silica gel (average 100 mesh). This solvent remains free of impurities for months when stored in dark bottles.

2. Dichloromethane (C.P.). This solvent is purified by passage through silica gel. The effectiveness of purification is determined by measuring light absorption at 240 mμ in the Beckman spectrophotometer where the purified solvent should show no significant absorption. It remains free of impurities for many months at room temperature.

3. Concentrated HCl, (C.P.).

4. Potassium hydroxide, 5% (w/v aqueous solution).

5. Ethyl alcohol. The alcohol is purified by the addition of 7 g. silver nitrate and 15 g. potassium hydroxide, separately (each dissolved in 100 ml. ethanol) to 4 l. of absolute ethanol. This is mixed and after storage in the dark for 16 to 24 hours, the alcohol is transferred by decantation to a distillation flask and distilled through a Vigreaux column. Discard the first 700 ml. and the last 300 ml. portions. The effectiveness of purification is determined by preparing a 1% solution of purified m-dinitrobenzene in the purified ethyl alcohol, and observing for absence of red color on addition of an equal volume of 5 N potassium hydroxide.

6. m-Dinitrobenzene (1% w/v alcohol solution). Reagent grade m-dinitrobenzene is purified by dissolving 30 g. in 1 l. of absolute alcohol. Warm to 40°C.; add 100 ml. of 2N NaOH, and after standing for 5 minutes, cool to 25°C. and add 3 l. of H_2O. The precipitated m-dinitrobenzene is collected on a Buchner funnel, washed several times with water, and recrystallized 3 to 4 times from 100 to 150 ml. volumes of absolute alcohol. This material must crystallize in nearly colorless needles, M.P. 90.5 to 91°C. One gram of recrystallized m-dinitrobenzene is dissolved in 100 ml. redistilled absolute ethyl alcohol and stored in a brown bottle.

NOTE: A highly purified m-dinitrobenzene requiring no further purification may be obtained from Sigma Chemical Company, St. Louis, Missouri.

7. Benzyl trimethyl ammonium methoxide (40% in methanol). Obtained from K and K Laboratories, Jamaica, N.Y., or Matheson, Coleman and Bell, East Rutherford, N.J.

8. Ethyl alcohol, 50% (v/v). This concentration of ethyl alcohol may be prepared by the appropriate dilution of a good commercial grade absolute alcohol with an equal volume of distilled water.

9. *Dehydroepiandrosterone standard.* Dissolve 50 mg. dehydroepiandrosterone, (Mann Research Laboratories, New York, N.Y.) in redistilled ethyl alcohol and make to 50 ml. Dilute 10 ml. of this solution to 100 ml. with alcohol, equivalent to 100 μg. dehydroepiandrosterone per ml.

Procedure

Pipet 5 ml. of urine into a 40-ml. graduated ground glass-stoppered conical centrifuge tube and add 0.5 ml. of concentrated HCl. Place the tubes in a boiling water bath for 20 minutes, and cover the top of each tube with a marble.

NOTE: All glassware used in this procedure must be scrupulously cleaned, first with an alkaline detergent and then by soaking in concentrated sulfuric acid prior to repeated rinsings with distilled water.

NOTE: If the urine is kept refrigerated no preservative need be added.

NOTE: For maximum yield with 10% (v/v) HCl, it is necessary to heat the tubes in the boiling water bath for at least 15 minutes. They should not be heated for more than 25 minutes because excessive heating will lead to destruction of the steroids and formation of chromogenic substances.

Remove the tubes, cool, and extract with 25 ml. of the mixture of petroleum ether-benzene (1 : 1). Shake 15–20 seconds, and after separation of the solvents, aspirate the urine from the lower layer by inserting a capillary pipet through the solvent layer. Wash the solvent with 1/15 volume of 5% KOH and discard the KOH washing by aspiration. Wash the solvent twice with 1/10 volume water, and discard in the same manner as the KOH.

NOTE: The alkali wash, in addition to removing the estrogen, also removes some acidic chromogen and chromogenic substances.

NOTE: It is necessary to completely remove the KOH by efficient aspiration and adequate water washes, since residual alkali will interfere with color development.

On the basis of the anticipated level of 17-ketosteroid in the sample a 5, 10, 15, or 20-ml. aliquot of the solvent is pipetted into a 30-ml., 19 × 150 mm. diameter tube, and then taken to dryness in a water bath (40–45°C.) under a stream of air. The extract is washed from the wall of the tube into the tip with ethanol.

To the dried residue, add 0.1 ml. of ethanolic *m*-dinitrobenzene, and

154   R. E. PETERSON

mix to completely dissolve the residue. Add 0.2 ml. of benzyl trimethyl-ammonium methoxide, and after mixing, incubate in the dark at 25°C. for 90 minutes.

NOTE: The temperature of incubation is not critical, however, at a lower temperature the color develops at a slower rate, whereas at higher temperatures color development will be more rapid but of less intensity. At 25° maximal color development occurs in 80–90 minutes. The incubation may be carried out in subdued light but away from bright light.

Reagent blanks are prepared in duplicate by adding 0.1 ml. of ethanolic m-dinitrobenzene and 0.2 ml. alkali to clean tubes. Standards are prepared in duplicate by adding 0.2 ml. of the 100 μg. ml. dehydro-epiandrosterone to similar tubes. The solvent is removed by evapora-tion and 0.1 ml. of ethanolic m-dinitrobenzene, and 0.2 ml. of alkali are added.

After 90 minutes incubation of the blank, standard, and unknown, add 3.0 ml. of 50% ethanol in water to each tube and mix. Add 3.0 ml. of dichloromethane and mix vigorously for 10 seconds.

Let the sample stand in the dark until the two layers separate and the bottom layer becomes crystal clear. Measure the absorbance (optical density) of the colored product (lower layer) against a water blank at 520 mμ in a Coleman, Jr. spectrophotometer with the cuvette tube raised by insertion of a 3 mm. high cork support in the bottom of the cuvette carrier.

NOTE: With suitably purified dichloromethane there should be less than 5% color fading per hour. The molar absorbance index for dehydroepiandrosterone at 520 mμ is 20,000.

Calculations

$$\frac{DU}{DS} \times 0.02 \times \frac{TVU}{UA} = \text{mg. ketosteroid per day}$$

DU = optical density unknown—optical density reagent blank

DS = optical density standard—optical density reagent blank

0.02 = mg. standard (dehydroepiandrosterone)

TVU = total 24 hr. urine volume

UA = urine aliquot used, ml. (5, 10, 15, or 20-ml. aliquot solvent equivalent to 1, 2, 3, or 4 ml. urine).

Discussion

Of the three major urinary 17-ketosteroids, two (androsterone and etiocholanolone) occur predominately as water-soluble conjugates with glucoronic acid, whereas dehydroepiandrosterone occurs primarily as the sulfate ester (3, 4). Both the ketosteroid sulfates and glucuronides can be cleaved with strong acids at elevated temperatures (5); however, this procedure, although suitable for routine assays, has two disadvantages. First, pigments and nonsteroidal chromogens are formed which are difficult to remove and may interfere with the colorimetric determination. Also, strong acids and elevated temperatures may destroy certain of the steroids and produce alterations in the molecule. The unsaturated steroid (dehydroepiandrosterone) sulfates are altered more readily under these conditions than the saturated sulfates. These transformations are products of dehydration, halogenation, and rearrangement (3).

Several procedures have recently been introduced for the differential hydrolysis of the respective sulfates and glucuronides. β-Glucuronidase will liberate 60–70% of the 17-ketosteroids released by boiling the urine for 15 minutes with 10–15% by volume HCl (6, 7, 8). Androsterone and etiocholanolone account for most of the neutral Zimmerman chromogens after glucuronidase hydrolysis. β-Glucuronidase will not liberate dehydroepiandrosterone and other steroids that are excreted in part as sulfates. Both the saturated and unsaturated sulfates can be hydrolyzed without transformations in structure under acidic conditions at room temperature when continuous extraction with ether is employed (9). Also, the procedure of solvolysis without continuous extraction will cleave the sulfates and produce little or no transformation of the 17-ketosteroids (10). Other procedures have been reported for hydrolysis of the steroid sulfates (11, 12, 13, 14, 15, 16, 17), and with all of these methods it has been demonstrated that temperature, pH, anion content, and time of hydrolysis can alter the nature of the products formed.

At present no single hydrolytic procedure is adequate for all urinary neutral 17-ketosteroids conjugates, and to achieve maximal cleavage of the water-soluble conjugates, a combination of procedures must be employed (8). Although the procedure of hydrolysis of the urine by boiling under strongly acid conditions is known to cause considerable alteration in the steroid molecule, at present this would appear to be the procedure of choice for the routine determination of the Zimmerman reacting steroid chromogens. Although slightly lower values are obtained with this procedure when compared with methods that utilize a combination of hydrolytic procedures, certain of the transformation products produced by this procedure probably retain their reactivity with the Zimmerman reagent.

A number of organic solvents may be used for extraction of the free 17-ketosteroids from urine after hydrolysis. The proper solvent will enhance the selectivity of the method, and readily extract the 17-ketosteroids and only a minimal amount of impurities, (pigments, chromogenic substances, and other materials that may interfere with the colorimetric assay). The more polar organic solvents, such as chloroform, dichloromethane, ethylene dichloride, and ethyl acetate, extract a large quantity of these interfering substances and thus may yield falsely high 17-ketosteroid values. The petroleum ethers are excellent solvents for the 17-ketosteroids with only two oxygen functions, and they extract only a minimum amount of interfering substances. However, these ethers show an unfavorable partition coefficient for 17-ketosteroids with more than two oxygen functions (8). Steroids with three oxygen functions make up 10 to 20% of the total neutral urinary Zimmerman-reacting chromogens (18). Ethyl ether, benzene, and carbon tetrachloride have been the solvents most widely used for extraction of the urinary 17-ketosteroids. The former is highly inflammable and prone to develop peroxides, whereas the latter two have relatively high toxicity hazards. Each of these three solvents will readily remove the 17-ketosteroids with either two or three oxygen functions. Ethyl ether removes more of the urinary pigments but fewer urinary chromogens, whereas benzene and carbon tetrachloride remove more of the urinary chromogens (8). A mixture of equal parts of benzene and petroleum ether will extract the ketosteroids quantitatively and remove fewer urinary chromogens and pigments. Also, once adequately purified the solvent mixture remains stable for many months at room temperature.

Except for the saturated hydrocarbons (petroleum ethers), the more polar solvents will remove the free 17-ketosteroids from the urine quantitatively during a single extraction when the ratio of solvent to aqueous phase is greater than 3 to 1. In the method described a solvent to urine ratio of 5 to 1 is used, because unwieldy emulsions are not encountered with this combination. Therefore, continuous solvent extraction would not appear to offer any advantage. Simultaneous hydrolysis and extraction does not increase the yield of 17-ketosteroids,

(11, 19). Even though a significant purification is obtained by utilizing a solvent relatively selective for the urinary 17-ketosteroids and an alkaline wash of the solvent extract, a considerable amount of nonketonic steroidal and ketonic and nonetonic nonsteroidal material remains in the crude neutral fraction and may interfere with the colorimetric assay.

Additional purification of the urine extract may be achieved by several procedures, none of which can be performed easily and routinely. The alkaline wash, in addition to removing much pigment, also removes organic acids and the phenolic 17-ketosteroids (estrogens).

Although many modifications of the original method of Zimmerman have been employed, most laboratories use one of two procedures, either in their original or slightly modified form. Both the Callow, (20), and the Holtorff and Koch (21) methods have in common the incubation of the alcoholic solution of 17-ketosteroids with a strongly alkaline solution of m-dinitrobenzene at a constant temperature in the dark. The two methods differ in the concentrations of alcohol and KOH used in the reaction mixture. In the Callow method the alcohol concentration is 100%, and the concentration of alkali 0.83 N, whereas in the Holtorff-Koch method the alcohol concentration is 50%, and the KOH 1.67 N. Both procedures give a red-purple color with absorption maxama at 520 mμ, and brownish pigments absorbing strongly at 400 to 440 mμ (8, 22). This combination of pigments develops when either pure 17-ketosteroids or crude urine extracts are incubated with alcohol, strong alkali, and m-dinitrobenzene. However, urine extracts yield a much larger amount of the brown pigments than pure 17-ketosteroids. Ketone groups at other positions of the steroid nucleus or in the side chain also may yield brown-colored products. It is these other ketonic steroids, nonsteroidal ketones, and nonketonic chromogens that complicate estimations of urinary 17-ketosteroids.

Several procedures have been utilized for correcting for the nonspecific brown pigments. The procedure described herein represents a relatively simple technique for removing the brown pigments from the red chromophore prior to spectrophotometry. After incubation the reaction mixture is diluted with 50% alcohol, and the red chromophore is extracted with dichloromethane. The brown pigments are retained nearly completely in the upper aqueous-alcohol phase (8). A similar procedure using chloroform was used by Cahen and Salter (23) and more recently other modifications have been reported (24, 25, 26, 27, 28). After investigation of many different solvents two were found to be suitable: dichloromethane and ethylene dichloride. Most other solvents incompletely extract the red color, (24, 26), give a decreased intensity of color, (23, 24, 26, 28), cause color fading, or produce a turbid solvent layer. When an inorganic alkali such as KOH is used to develop the Zimmerman chromogen the dichloromethane also shows a faint turbidity, requiring the addition of alcohol, (8), or poly-ethyleneglycol (28) to the dichloromethane, or centrifugation (27). Substitution of the organic base benzyl trimethyl ammonium meth-oxide for the KOH yields a crystal clear dichloromethane layer, which may be read directly in the spectrophotometer without removal of the aqueous-alcohol phase. Ethylene dichloride produces a crystal clear solution, but the purified solvent is unstable and requires fre-quent repurification. Dichloromethane, once adequately purified, re-mains stable for long periods of time at room temperature.

The contribution of the urinary pigments may be eliminated by assaying a blank in which alcohol is substituted for m-dinitrobenzene, as recommended by Holtorff and Koch (21). However, the dichloro-methane extraction procedure effectively removes most urine pigments, and it is not necessary to run a blank. Treatment of an ethylene dichlo-ride extract with solid NaOH pellets was reported to remove a large amount of the pigments formed during acid hydrolysis (29). However, if a less polar solvent is used, fewer pigments are extracted from the urine. The addition of formalin to the urine prior to hydrolysis will effectively decrease the formation of interfering pigments (30). How-ever, neither of these two procedures offers any advantage provided a relatively nonpolar solvent (benzene-petroleum ether) is used for extraction of the urine, and the dichloromethane extraction is used prior to spectrophotometric assay.

Urines collected during the latter part of pregnancy, urines hy-drolyzed with β-glucuronidase, or urines collected from subjects treated with corticosteroids or corticotropin may contain large quan-tities of non-17-ketosteroids (C_3, C_{11}, C_{20} ketones). These ketonic steroids give a significant amount of brown color in the Zimmerman reaction. Also, if present in high concentration they may reduce the reactivity of the 17-ketosteroids and lead to incomplete recovery of 17-ketosteroids in the crude neutral extract (31). Extraction of the hydrolyzed urine with a selective solvent will diminish the amount of interference due to non-17-ketosteroids and improve the accuracy of their measurement. Dichloromethane extraction of the incubated resi-due will effectively remove most of the brown pigments but will not enhance the recovery of the netural 17-ketosteroids (8). Except for urines that contain a high normal or elevated titer of 17-ketosteroids, incubation of the crude extract with the Zimmerman reagents usually yields substances that absorb with greater intensity at 420 mμ than at 520 mμ (8, 22, 23, 24, 25, 26, 27, 28, 32, 33). This may be even more apparent in urine from patients with various diseases. A large part of this chromogenic material that absorbs maximally at 420 mμ is non-ketonic and can be separated from the ketonic material by the aid of the Girard reagent T, (8, 32, 33, 34). The use of a relatively nonpolar solvent for extraction of the urine and dichloromethane extraction of the incubated residue will markedly diminish the interference from the nonketonic chromogens (8).

"Color correction equations" have been used as a mathematical device for dealing with the interfering pigments and chromogens contributing in part to the absorbance of light at the 520 mμ wavelength. For a correction based on readings at 420 mμ and 520 mμ, the absorbance ratio for the 17-ketosteroids and for the interfering substances at the two wavelengths must be known. It is possible to determine the 520 mμ vessus 420 mμ absorbance ratio for pure 17-ketosteroid; however, it is not readily feasible to obtain such a ratio for the interfering chromogens. Usually the ratio has been determined on the nonketonic fraction of crude urine extracts after Girard T fractionation (34, 35, 36). However, depending on the urine specimen assayed, the concentration used, the method of hydrolysis, and the solvent used for extraction, the ratio will show great variability (33). In addition, destruction of some chromogenic material occurs during the Girard purification (33). Also, ketonic fractions of normal urines or urine extracts that are free of 17-ketosteroids do not give the same absorption spectra as the pure 17-ketosteroids (32).

By using Allen's correction equation (37), a corrected optical density is obtained from readings at three wavelengths. The correction is valid provided that the absorbance of the nonspecific pigments and chromogens at the three wavelengths is linear. The absorption of the nonketonic chromogens is essentially a linear function of wavelength over the range of 460 to 600 mμ (32), however, a significant fraction of the interfering chromogens showing absorption at 520 mμ is in the ketonic residue. Urine extracts that are essentially free of 17-ketosteroids usually do not show a linear absorption over the range of 460 to 600 mμ. Thus, the Allen correction is not entirely valid.

The organic base benzyl trimethyl ammonium methoxide and
m-dinitrobenzene was introduced in the colorimetric detection of various ketones by Pratt, E. L. (38), and recommended for the determination of 17-ketosteroids by Bongiovanni *et al.* (39). It offers the
following advantages over the conventional aqueous or alcoholic
solution of KOH: (1) It shows a lower reagent blank reading than
2.5 N KOH in alcohol, (2) it yields approximately the same intensity
of color with the three major neutral urinary 17-ketosteroids (etiocholanolone, androsterone, dehydroepiandrosterone), (3) it produces
approximately 100% more color than the inorganic alkalis; although
this may vary somewhat with different batches of the base, (4) it
causes very much less color fading than with inorganic alkali (40), and
(5) with the use of dichloromethane for the extraction of the Zimmerman-reacting red chromophore, the solvent layer remains crystal
clear, rather than opalescent.

REFERENCES

1. von Bitto, B., Ueber eine Reaction der Aldehyde und Ketone mit Arom. Nitroverbindungen. *Ann. Chem.* **269**, 377 (1892).
2. Zimmermann, W., Eine Farbreaktion der Sexualhormone und ihre Unwendung zur quantitativen colorimetrischen Bestimmung. *Z. Physiol. Chem.* **233**, 257–264 (1935).
3. Lieberman, S., Mond, B., and Smyles, E., Hydrolysis of urinary ketosteroid conjugates. *Recent Progr. Hormone Res.* **11**, 113–134 (1954).
4. Kellie, A. E., and Wade, A. P., Steroid conjugates. (1) The separation of urinary 17-ketosteroid glucuronides and sulphates, and their composition in normal individuals. *Acta Endocrinol.* **23**, 357–370 (1956).
5. Buehler, H. J., Katzman, P. A., and Doisy, E. A., Hydrolysis of conjugates of neutral ketosteroids. *Proc. Soc. Exptl. Biol. Med.* **78**, 3–8 (1951).
6. Buehler, H. J., Katzman, P. A., Doisy, P. P., and Doisy, E. A., Hydrolysis of conjugated steroids by bacterial glucuronidase. *Proc. Soc. Exptl. Biol. Med.* **72**, 297 (1949).
7. Cohen, S. L., The hydrolysis of steroid glucuronides with calf spleen glucuronidase. *J. Biol. Chem.* **192**, 147–160 (1951).
8. Peterson, R. E., and Pierce, C. E., "Methodology of Urinary 17-Ketosteroids, Lipids and the Steroid Hormones in Clinical Medicine," Sunderman and Sunderman, ed. Lippincott, Philadelphia, Pennsylvania, 1960.
9. Lieberman, S., and Dobriner, K., Steroid excretion in health and disease; chemical aspects. *Recent Progr. Hormone Res.* **3**, 71–101 (1948).
10. Burstein, S., Hydrolysis of ketosteroid hydrogen sulfates by solvolysis procedures. *J. Biol. Chem.* **233**, 331–335 (1958).
11. Talbot, N. B., Ryan, J., and Wolfe, J. K., The determination of sodium dehydroisoandrosterone sulfate in water or urine, *J. Biol. Chem.* **148**, 593–602 (1943).

12. Barton, D. H. R., and Klyne, W., Identification of "17-ketosteroid II" as i-androsten-6-ol-one. *Nature* **162**, 493–494 (1948).
13. Bitman, J., and Cohen, S. L., Hydrolysis of urinary conjugated 17-ketosteroids by acetate buffer. *J. Biol. Chem.* **191**, 351–363 (1951).
14. Dingemanse, E., and Huis in't Veld. L. G., Isolation of i-androstan-6-ol-17-one from urine of patients with adrenal neoplasm. *J. Biol. Chem.* **195**, 827–835 (1952).
15. Cohen, S. L., and Oneson, I. B., The conjugated steroids. IV. The hydrolysis of ketosteroid sulfates. *J. Biol. Chem.* **204**, 245–256 (1953).
16. Teich, S., Rogers, J., Lieberman, S., Engel, L. L., and Davis, J. W., The origin of 3,5-cycloandrostan-6β-ol-17-one (i-androsten-6β-ol-17 one) in urinary extracts. *J. Am. Chem. Soc.* **75**, 2523–2525 (1953).
17. Fotherby, K., A method for the estimation of dehydroepiandrosterone in urine. *Biochem. J.* **73**, 339–343 (1959).
18. Dobriner, K., Studies in steroid metabolism. XIX. The α-ketosteroid excretion pattern in normal males. *J. Clin. Invest.* **32**, 940–949 (1953).
19. Jensen, C. C., and Totterman, L. E., Hydrolysis of urinary neutral 17-ketosteroid conjugates, I. Comparison of various procedures. *Acta Endocrinol.* **10**, 221–232 (1952).
20. Callow, N. H., Callow, R. K., and Emmens, C. W., Colorimetric determination of substances containing the grouping—CH₂CO—in urine extracts as an indication of androgen content. *Biochem. J.* **32**, 1312–1331 (1938).
21. Holtorff, A. F., and Koch, F. C., The colorimetric estimation of 17-ketosteroids and their application to urine extracts. *J. Biol Chem.* **135**, 377–392 (1940).
22. McCullagh, D. R., Schneider, I., and Emery, F., Spectrophotometric studies of the *m*-dinitrobenzene reaction with sex steroid. *Endocrinology* **27**, 71–78 (1940).
23. Cahen, R. L., and Salter, W. T., Urinary 17-ketosteroids in metabolism. I. Standardized chemical estimation. *J. Biol. Chem.* **152**, 489–499 (1944).
24. Zimmermann, W., Anton, H. U., and Pontius, D., Die Eliminerung storenden Chromogene bei der Bestimmung der 17-ketosteroids mit *m*-dinitrobenzol. *Z. Physiol. Chem.* **289**, 91–102 (1952).
25. Masuda, M., and Thuline, H. C., An improved method for determination of urinary 17-ketosteroids. *J. Clin. Endocrinol.* **13**, 581–586 (1953).
26. Werbin, H., and Ong. Siew., Improved colorimetric determination of urinary 17-ketosteroids. *Anal. Chem.* **26**, 762–764 (1954).
27. Ware, A. G., Demetrion, J. A., Notrica, S., Seacy, R., Walberg, C. and Cox, F., Elimination of interfering chromogens in the Zimmermann reaction for measuring 17-ketosteroids. *Clin. Chem.* **5**, 479–487 (1959).
28. Rappaport, F., Fischl, J. and Pinto, N., A rapid method for the estimation of urinary 17-ketosteroids. *Clin. Chem.* **6**, 16–22 (1960).
29. Drekter, I. J., Heisler, A., Scism, G. R., Stern, S., Pearson, S., and McGavack, T. H., The determination of urinary steroids; I. The preparation of pigment-free extracts and a simplified procedure for the estimation of total 17-ketosteroids. *J. Clin. Endocrinol.* **12**, 55–65 (1952).
30. Antunes, L. N., Interfering chromogens in the 17-ketosteroid determination. *J. Clin. Endocrinol.* **16**, 1125–1126 (1956).

162

31. Sloan, C. H., and Lowery, G. H., The elimination of common error in the determination of total 17-ketosteroids by the *m*-dinitrobenzene reaction. *Endocrinology* 48, 384–390 (1951).
32. Chang, E., and Slaunwhite, W. R., A comparison of urinary Zimmermann chromogens by Girard fractionation and by Allen's method of correction. *J. Clin. Endocrinol.* 15, 767–775 (1955).
33. Schedl, H. P., Bean, W. B., Stevenson, B. M., and Schumacher, E. R., Correction for color differences between standards and urine extracts and their ketonic fractions in the Callow-Zimmermann reaction. *J. Lab. Clin. Med.* 45, 191–214 (1955).
34. Pincus, G., and Perlman, W. H., Fractionation of neutral urinary steroids. *Endocrinology* 29, 413–424 (1941).
35. Fraser, R. W., Forbes, A. P., Albright, F., Sulkowitch, H., and Reifeinstein, E., Colorimetric assay of 17-ketosteroids in urine. *J. Clin. Endocrinol.* 1, 234–256 (1941).
36. Beher, W. T., and Gaebler, O. H., Determination of neutral 17-ketosteroids in urine. *Anal. Chem.* 23, 118–123 (1951).
37. Allen, W. M., A simple method for analyzing complicated absorption curves of use in the colorimetric determination of urinary steroids. *J. Clin. Endocrinol.* 10, 71–83 (1950).
38. Pratt, E. L., Colorimetric method for estimation of digitoxin. *Anal. Chem.* 24, 1324 (1952).
39. Bongiovanni, A. M., Eberlein, W. R., and Thomas, P. Z., Use of an organic base in the Zimmermann reaction. *J. Clin. Endocrinol.* 17, 331–332 (1957).
40. Klendshoj, N. C., Feldstein, M., and Sprague, A., Determination of 17-ketosteroid in urine. *J. Clin. Endocrinol.* 13, 922–940 (1953).

LACTIC DEHYDROGENASE*

Submitted by: GEORGE N. BOWERS, JR., Departments of Pathology and Medicine,
Hartford Hospital, Hartford, Connecticut

Checked by: YASUO TAKENAKA, Grace-New Haven Community Hospital, New
Haven, Connecticut. (Present address: Children's Hospital, 226
North Kuakini Street, Honolulu 17, Hawaii)

Introduction

Several enzymes catalyze the removal of hydrogen from lactate,
however, only those requiring the coenzyme nicotinamide-adenine
dinucleotide (NAD, formerly DPN) as the hydrogen acceptor, are
classified as lactic dehydrogenases (LDH). These enzymes have been
highly purified and crystallized from mammalian heart, liver, and
skeletal muscles, as well as from yeasts and bacteria (1). During the
1930's, Warburg and Christian developed the spectrophotometric technique for rapidly following the activity of an NAD-dependent enzyme,
thus permitting accurate measurements to be performed with great
ease (2).

The nicotinamide-adenine dinucleotide dependent enzymes of glycolysis were extensively studied for their relation to tumor metabolism
by von Euler's (3) group as well as Warburg and associates. The
latter investigators reported on findings of elevated serum LDH
activity in tumor-bearing rats (4), a finding substantiated by many
others (5, 6, 7). There have been numerous reports of high LDH
activity occurring in human sera in association with carcinomas
(8, 9, 10, 11, 12), leukemia, and megaloblastic anemias (13, 14). However, it has been the finding of consistently elevated activities following acute myocardial infarction, first reported in 1955 by Wroblewski
and LaDue (15) which is most useful clinically and is therefore most
frequently used.

Methods

Lactic dehydrogenase activity has been measured by manometric,
chromatographic, flurometric, colorimetric, and spectrophotometric

* Based on the methods of Hill (9), Wroblewski and LaDue (15), and Henry
et al. (20).

163

techniques. Colorimetric procedures have employed either the Thunberg methylene blue principle or the more commonly used 2,4-dinitrophenylhydrazine method. The spectrophotometric methods depend upon the marked absorbancy of $NADH_2$ at 340 mμ versus the insignificant absorbancy of NAD at 340 mμ. The assay may be performed equally well starting with lactate-NAD or pyruvate-$NADH_2$ as the substrate coenzyme mixture. The latter procedure is described below and was chosen for several rather arbitrary reasons to be enumerated in the discussion.

Principle

The oxidation of $NADH_2$ to NAD is measured spectrophotometrically at 340 mμ as pyruvate is enzymatically reduced to lactate.

$$
\begin{array}{ccccc}
\underset{|}{\overset{CH_3}{\overset{|}{C}}}= O & + \ NADH_2 & \underset{\xrightarrow{\quad}}{\xrightarrow{\text{LDH}}} & NAD \ + & \underset{|}{\overset{CH_3}{\overset{|}{C}}HOH} \\
| & & & & | \\
COOH & & & & COOH \\
\text{Pyruvate} & & & & \text{Lactate}
\end{array}
$$

Since a stoichiometric relationship is generally assumed for this reaction, the decreasing absorbance of $NADH_2$ is considered to be equivalent to the consumption of pyruvate. Therefore, the basic unit as defined by the Joint Sub-Commission on Clinical Enzyme Units of the International Union of Biochemistry and the International Union of Pure and Applied Chemistry, Munich 1959, (16), should be equal to a micromole of pyruvate transformed per minute per milliliter of serum at 37.5°C. However, the unit which is defined as a change of 0.001 absorbance per minute per milliliter of serum at 25°, 32°, or 37.5°C. is more widely used.

Reagents

A. Stock Reagents:[1]

1. *Tris (hydroxymethyl) aminomethane* ("Tris") should be of the highest purity, recrystallized reagent grade suitable for enzyme determinations, i.e. Sigma 121.[1]

[1] Reagents were obtained from Nutritional Biochemical Co., Cleveland and Sigma Chemical Co., St. Louis, Missouri.

2. *Dihydronicotinamide-adenine dinucleotide* ($NADH_2$), highest grade available stored in desiccator in the dark, i.e., Sigma 98–100% pure.

3. *Sodium pyruvate*, highest grade analytical reagent.

4. *Potassium dichromate*, analytical reagent grade.

B. *TRIS-NADH₂ Mixture* (directions are for the manufacture of 100 mixed reagent tubes). Stable for 10 days.

1. Dissolve 3.0 g. Tris in 260 ml. of distilled-deionized water.

2. Adjust pH to 7.45 ± 0.05 with 6 *N* HCl.

3. Add 50 mg. $NADH_2$ and mix well. The absorbance should be 1.30–1.40, otherwise, more or less $NADH_2$ should be used.

4. Pipette 2.75 ml. of this Tris-$NADH_2$ mixture into clean dry tubes, cap to prevent evaporation and store at 4°C.

C. *TRIS-PYRUVATE Mixture*; To 100 ml. H_2O add 1.2 g. Tris and 150 mg. of Na pyruvate, mix well, adjust the pH with HCl to 7.45 ± 0.05 and store at 4°C. Make pyruvate *each* week.

Procedure

1. Add 0.050 ml. of nonhemolyzed fresh serum to 2.75 ml. of Tris-$NADH_2$ reagent, mix well and incubate at 37.5°C. for at least 20 minutes to remove serum pyruvate. For serums with high activity use a 1 to 10 dilution of serum in normal saline.

2. Initiate the LDH reaction by adding 0.20 ml. of warmed (37.5°C.) Tris-pyruvate reagent. Mix quickly by several inversions and transfer to a $1 \times 1 \times 3.5$ cm. glass cuvette which has been warmed to 37.5°C. in the water bath (light path, 1.00 cm.).

3. Wipe the cuvette dry and place in the heated 37.5°C. reading compartment of a spectrophotometer capable of producing readings at 340 mμ.[2]

4. Record the absorbance at 15- or 30-second intervals for 2 to 3 minutes, against a $K_2Cr_2O_7$ reference. The absorbance of the dichromate reference is adjusted to 0.500–0.600 so that the initial absorbance of the tests fall at 0.500 ± 0.100.

---

[2] Beckman, B, DU, DK-1; Zeiss; Perkin-Elmer; and Spectronic 20 spectrophotometers have been used successfully as has a simple UV photometer (18).

Calculations

A. The basic international unit of micromoles per milliliter per minute at 37.5°C. is:

$$\text{LDH}_{37.5°C.} = \frac{\Delta A \text{ per minute}}{6.22} \times \frac{V}{D}$$

where ΔA is the change in absorbance, 6.22 is the absorbance of one micromole of NADH_2 per milliliter at 340 mμ for 1 cm. light path (17), V is the total volume of the reaction mixture in milliliters, and D is the serum sample size in milliliters.

B. The unit for a total volume of 3 ml. as defined by change of absorbance is:

$$\text{LDH}_{37.5°C.} = \frac{\Delta A \text{ per minute}}{\text{ml. of serum}} \times 1000$$

Discussion

The spectrophotometric method for measuring lactic dehydrogenase activity was chosen in preference to the equally popular 2,4-dinitro-phenylhydrazine colorimetric procedure for several reasons. The most important is that the method as outlined here using premixed reagent has proved to be rapid, precise, and economical. Although initially it may require more effort to set up the spectrophotometric method, the ability to work efficiently at 340 mμ permits a wide variety of clinically useful coenzyme-linked-dehydrogenase enzymes or their substrates to be determined. Since there are well over 60 of these dehydrogenases recognized in various metabolic steps, we can expect an ever increasing number of investigations to delineate clear cut clinical applications for these enzymes. As an example, this laboratory presently employs the lactic, malic, isocitric, alcohol, and glucose-6-phosphate dehydrogenase systems to measure the serum activity of LDH, MDH, SGPT, SGOT, ICD, blood alcohol concentration, and red cell content of G-6-PO$_4$ dehydrogenase respectively.

Spectrophotometers capable of efficient operation in the ultraviolet wave lengths must be considered basic operational equipment for modern hospital chemistry laboratories, not only to reach 340 mμ for these coenzyme-linked reactions, but also for identification and quantification of such important compounds as the barbituric acids and phenothiazides. Since excellent instruments, the operational characteristics of which have been fully proven, are readily available at costs well within the budget of nearly all hospital laboratories, the choice of a spectrophotometric method may prove to be simpler and more advantageous in the long run.

The problem of which direction of the enzyme reaction represents the best way to perform the LDH assay has been the subject of controversy (19, 20).

The advantages of utilizing the pyruvate to lactate reaction as described in this method are as follows:

1. At body pH and temperature the reaction rate of pyruvate to lactate is three times as fast as that of the reverse reaction, lactate to pyruvate. Therefore, the activity of smaller serum samples (20–100 μl.) may be accurately measured or shorter periods of observation time in the spectrophotometer may be used.

2. $NADH_2$ is stable in alkaline solution, thus permitting it to be incorporated into the Tris buffered reagent. NAD on the other hand, decomposes in alkaline solutions and mixtures must be lyophilized for stability.

3. $NADH_2$ in 0.1 M Tris buffer at 7.45 rather than the usual phosphate buffers, has shown less than a 4% decrease in absorbance per week when stored at 4°C. This stability of the $NADH_2$ in Tris has permitted the amino acids and dehydrogenases for the transaminase reactions to be incorporated into premixed transaminase reagents.[3]

4. Since one of the major criticisms of the spectrophotometric methods is the cost of reagents, it is important to note that at present reagent prices, the cost per test for the pyruvate to lactate reaction is approximately 1.5 cents. However, the higher concentrations of DPN required in the reverse reaction, raise the costs to approximately 10 cents per test.

5. Perhaps what seems to be a disadvantage of the pyruvate to lactate reaction, the 20 minutes waiting period during the incubation to exhaust endogenous serum pyruvate, can be construed as an advantage for it permits the addition of an optimal concentration of pyruvate short of that amount which would cause substrate inhibition. The major difficulty encountered in over 2 years of daily use of this method has not been by inhibition from pyruvate but by an unidentified LDH inhibitor in some batches of new $NADH_2$. However, since

[3] In addition to the amounts of H_2O, Tris, and DPNH listed, a mixture of 100 transaminase tubes contains 5 gm. of either L-aspartate or L-alanine plus enough MDH or LDH to give a zero order reaction for activities up to 1000 absorbance units. The pH is again adjusted to 7.45 ± 0.05 with NaOH or HCl.

the report by Fawcett *et al.* (21) on LDH inhibitors developing with freezing of $NADH_2$ solutions, all reagents have been stored at 4°C. with no demonstrable loss of maximal LDH activity after 3 weeks at this temperature.

Proportionality

A proportional relationship between activity and serum sample size was observed for samples less than 0.2 ml. as shown in Fig. 1. Each point on this graph is the mean of five replicates.

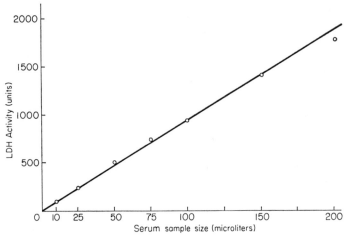

Fig. 1. Proportional relationship between LDH activity and serum sample size.

Reproducibility

The following are consecutive daily serum pool results performed independently by three technicians on two separate batches of reagents from March 1 to April 20.

Day	LDH	Day	LDH	Day	LDH	Day	LDH
1	460	10	480	19	480	28	460
2	480	11	470	20	450	29	430
3	440	12	500	21	440	30	445
4	470	13	430	22	430	31	460
5	440	14	460	23	480	32	470
6	485	15	460	24	490	33	460
7	465	16	485	25	485	34	450
8	480	17	445	26	460	35	460
9	460	18	440	27	485	36	440

Mean   460 units   Standard Deviation   ± 20 units

Stability of Serum LDH

While there is an immediate 40% increase in serum LDH over citrated plasma probably related to the liberation of enzymes from platelets during clotting, this enzyme is relatively stable for several days if held at 4°C. However, serum should be removed from the clot at the earliest possible time to avoid continuing LDH release from the red cells despite the absence of hemolysis.

Hemolysis

Hemolyzed serum must be rejected because of contamination by erythrocytes which contain 100 times the LDH activity of normal serum. Redrawing serum without hemolysis demonstrates the marked variation introduced by red cell lysis.

State of serum	Hemolyzed	Nonhemolyzed	Difference	% Normal
Gross hemolysis	1265	600	665	102
	1160	780	380	58
Moderate hemolysis	1400	1190	210	32
	915	720	195	30
Slight hemolysis	805	620	185	28
	660	510	150	23

Normal Values

Healthy hospital employees, ages 16 through 60, served as normal controls. All sera with even a hint of hemolysis were rejected. Since temperature is undoubtedly one of the greatest variables in establishing normal LDH activity, all results were obtained with close control of temperature at 37.5°C.

Sex	No.	Mean	SD	Mean ± 2SD	Mean ± 3SD
Female	27	535	63	410–660	345–725
Male	16	570	58	455–685	395–745

In comparison with the normal LDH values published by Henry et al. (20), which were also obtained with careful attention to temperature control, these normal LDH values are 5% higher at both the mean and the 90% confidence level. The slightly higher activity is

probably related to the use of Tris buffer since minimal inhibition of
LDH activity is noted with increasing molarity of the phosphate
buffers. For simplicity, enzyme activities in excess of 750 units are
considered abnormal while those of lesser activity are "normal." If a
temperature other than 37.5°C. is employed conversion factors (given
on the Temperature-Activity graph in Fig. 2) should be used.

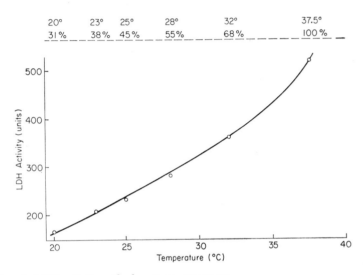

F IG. 2. LDH activity with changing temperature.

LDH Activity in Disease

The widespread distribution of LDH in human tissues frequently
makes its measurement a sensitive, albeit a nonspecific test. Elevations
are noted in some carcinomas, leukemia, severe hemolytic anemia,
megaloblastic anemias, hepatic disorders, ischemic vascular disease,
myocardial infarctions, and pulmonary emboli (22). Hamolsky and
Kaplan (23) have evaluated the laboratory diagnostic aids in acute
myocardial infarction and concluded that were it not for hemolysis
causing false LDH elevations, this would be the "enzyme-of-choice."
Recently, Smyrniotis et al. (24) have reported that gastric juice LDH
activity is useful in the diagnosis of gastric cancer, while Wacker and
Dorfman (25) have suggested that urinary LDH activity after dialysis
to remove inhibitors provides a screening method to detect cancer of
the kidney and bladder.

Molecular Heterogeneity of LDH

Serum LDH has been separated into several fractions (26) called by some, isozymes (27) and others isoenzymes (28). The techniques for fractionation have been mainly electrophoresis on paper, starch, agar, acrilamide, or cellulose acetate; however, column chromatography, NAD analogs, heat denaturation, and immunochemical reactions have also been employed (29). The clinical utility of measuring the LDH isozymes is rapidly becoming established as a result of numerous clinical investigative studies.

REFERENCES

1. Dixon, M., and Webb, E. C., "Enzymes." Academic Press Inc., New York (1958).
2. Warburg, O., and Christian, W., *Biochem. Z.* **286**, 61 (1936).
3. von Euler, H., and Schlenk, F., Pharmacology of cozymase. *Acta Physiol. Scand.* **1**, 19 (1940).
4. Warburg, O., and Christian, W., Garungsfermente in Blutserum von Tumor Ratten. *Biochem. Z.* **314**, 399 (1943).
5. Sister Lenta, M. Petra, and Sister Riehl, Agatha, Dehydrogenase studies of tissue from normal and Tumor-bearing mice, LDH's and MDH's. *Cancer Res.* **9**, 47 (1949).
6. Hsieh, K. M., Suntzeff, V., and Cowdry, E. V., Comparative study of serum LDH activity in mice with transplanted and induced tumors. *Cancer Res.* **16**, 237 (1956).
7. Bobansky, O., and Scholler, J., Comparison of PHI and LDH activities in plasma, liver and tumor tissue of tumor-bearing rats. *Cancer Res.* **16**, 894 (1956).
8. Hill, B. R., Borroughs, R., and Levi, C., Elevation of a serum component in neoplastic disease. *Cancer Res.* **14**, 513 (1954).
9. Hill, B. R., Some properties of serum lactic dehydrogenase. *Cancer Res.* **16**, 460-467 (1956).
10. Hess, B., and Gehn, E., Lactic acid dehydrogenase in human serum. *Klin. Wochschr.* **33**, 91 (1955).
11. Zimmerman, H. J., and Weinstein, H. G., Lactic dehydrogenase activity in human serum. *J. Lab. Clin. Med.* **48**, 607 (1956).
12. Moore, A. E., and Wroblewski, F., LDH produced in tissue culture of normal transformed and malignant human cell lines. *Proc. Soc. Exptl. Biol. Med.* **98**, 783 (1958).
13. Amelung, D., Serum enzyme determinations in pernicious anemia. *Deut. Med. Wochschr.* **85**, 1629-32 (1960).
14. Heller, P., Weinstein, H. G., West, M., and Zimmerman, H. J., Glycolytic, citric acid cycle and hexos emonophosphate shunt enzymes of plasma and erythrocytes in megaloblastic anemia. *J. Lab. Clin. Med.* **55**, 425 (1960).
15. Wroblewski, F., and LaDue, J. S., Lactic dehydrogenase activity in blood. *Proc. Soc. Exptl. Biol. Med.* **90**, 210 (1955).

172 G. N. BOWERS, JR.

16. Freeman, M. E., Clinical enzyme unitage. *Clin. Chem.* **7**, 199 (1961).
17. Horecker, B. L., and Kornberg, A., The extinction coefficients of the related band of pyridine nucleotides. *J. Biol. Chem.* **175**, 385 (1948).
18. Thiers, R. E., Margoshes, M., and Vallee, B. L., Simple ultraviolet Photometer. *Anal. Chem.* **31**, 1258 (1959).
19. Thiers, R. E., and Vallee, B. L., Analytical considerations in serum enzyme determinations. *Ann. N. Y. Acad. Sci.* **75**, 215 (1958).
20. Henry, R. J., Chiamori, N., Golub, O. J., and Berkman, S., Revised spectrophotometric methods for the determination of glutamic-oxalacetic transaminase, glutamic-pyruvic transaminase and lactic acid dehydrogenase. *Tech. Bull. Registry Med. Technologists* **30**, 149-66 (1960); *Am. J. Clin. Path.* **34**, 381-98 (1960).
21. Fawcett, C. P., Ciotti, M. M., and Kaplan, N. O., Inhibition of dehydrogenase reactions by substance formed from reduced diphosphopyridien nucleotide. *Biochem. Biophys. Acta* **54**, 210 (1961).
22. Erickson, R. J., and Morales, D. R., Clinical use of lactic dehydrogenase. *New Engl. J. Med.* **265**, 478 and 531 (1961).
23. Hamolsky, M. W., and Kaplan, N. O., Measurement of enzymes in the diagnosis of acute myocardial infarction. *Circulation* **23**, 102 (1961).
24. Smyrniotis, F., Schenker, S., O'Donnell, J., and Schiff, L., Lactic dehydrogenase activity in gastric juice for diagnosis of gastric cancer. *Am. J. Digest. Diseases* **7**, 712-26 (1962).
25. Wacker, W. E. C., and Dorfman, L. E., Urinary lactic dehydrogenase activity, I. Screening method for detection of cancer of kidneys and bladder. *J. Am. Med. Assoc.* **181**, 927-8 (1962).
26. Vesell, E. S., and Bearn, A. G., Localization of lactic acid dehydrogenase activity in serum fractions. *Proc. Soc. Exptl. Biol. Med.* **94**, 96 (1957).
27. Markert, C. L., and Moller, F., Multiple forms of enzymes. Tissue, ontogenetic and species specific patterns. *Proc. Natl. Acad. Sci. U. S.* **45**, 753-63 (1959).
28. Wroblewski, F., Ross, C., and Gregory, K. F., Isoenzymes and myocardial. infarction. *New Engl. J. Med.* **263**, 531 (1960).
29. Wroblewski, F. (guest editor), Multiple molecular forms of enzymes. *Ann. N. Y. Acad. Sci.* **94** (1961).

GRAVIMETRIC DETERMINATION OF TOTAL LIPIDS IN BLOOD SERUM OR PLASMA*

Submitted by: Warren M. Sperry, Departments of Biochemistry, New York State Psychiatric Institute and the College of Physicians and Surgeons, Columbia University, New York, New York

Checked by: S. Lawrence Jacobs, Bio-Science Laboratories, Los Angeles, California

Joseph Amenta, Capt., M.C., U.S.A., Walter Reed Army Medical Center, Washington, D.C.

Total lipids have been determined in blood serum or plasma largely by indirect methods developed from the procedure of Bang (1), who oxidized the lipids with acid chromate and measured the chromate consumed by iodometry. Under his conditions, the oxidation was incomplete and an empirical factor had to be applied in calculating the results. Bloor (2) made the oustanding contribution of finding conditions under which oxidation is nearly quantitative. In his widely used method the lipids are saponified, and the unsaponifiable material and fatty acids are oxidized. The so-called "total lipids" are calculated from the cholesterol concentration and theoretical factors for the oxidation of the cholesterol-fatty acid mixture. Bragdon (3) applied a colorimetric modification of the oxidative method to the unsaponified lipids of the serum. In his procedure determinations of cholesterol and also of phosphatides are necessary so the corresponding oxidation factors may be applied. In these and other modifications of the oxidative method average molecular weights of the complex mixture of serum lipids must be assumed, and it must also be assumed that the lipids being analyzed are free of oxidizable nonlipid contaminants (see below).

To avoid the uncertainties of the oxidative methods, it would appear to be a simple matter to isolate the unmodified lipids and weigh them, but there has been a serious obstacle to the development of a reasonably precise method of this type. Until recently there has been no adequate means of removing nonlipid impurities from serum lipids

---

* Based on the methods of Folch *et al.* (6, 7, 8).

173

174 W. M. SPERRY

except by conventional dialysis which is too cumbersome for routine use. In most methods, lipids are "purified" by evaporating the crude extract to dryness, usually in ethanol-ether, and treating the residue with a less polar solvent, usually petroleum ether. It has been shown (4, 5) that such a "purification" is not adequate; considerable amounts of nonlipid contaminants, principally urea, are extracted by petroleum ether along with the lipids. This obstacle was overcome by the development of the elegant methods of Folch and his colleagues (6, 7, 8) for the purification of lipids, and the method here described was made possible by the availability of these procedures.[1]

Reagents

1. Methanol, reagent grade, distilled, residue-free. The methanol used in our laboratory is distilled through a 40-plate column, as are most of the solvents we use, but such high purification is certainly not necessary in this method.

2. Chloroform, reagent grade, residue-free. Distill over K_2CO_3 into a bottle containing enough ethanol to make the final concentration 1%.

3. Ethanol, distilled, residue-free.

4. Chloroform-methanol (2 + 1, v/v). The constituents are carefully measured in graduated cylinders, previously rinsed with small portions of the solvents.

5. Calcium chloride solution. Dissolve 40 mg. anhydrous $CaCl_2$ per 100 ml. water.

Apparatus

Lab Jack, Central Scientific Co.

Dri-Jar desiccators, Bethlehem Apparatus Company, Hellertown, Pa.

Semimicro balance of high precision.

Equipment for transfer, filtration, and drying of lipid solutions (cf. Fig. 2 (9). Corks instead of rubber stoppers are now used in making apparatus for transfer (C).

Glassware. All glassware to be used to contain lipid solutions is rinsed with small portions of chloroform-methanol and dried, if necessary, before use.

---

[1] The author is greatly indebted to Dr. Jordi Folch for information about these procedures before they were published.

Procedure

EXTRACTION

Into 8.3 ml. of methanol in a 25-ml. volumetric flask is pipetted 1.00 ml. of blood serum or plasma. The tip of the pipet is held against the neck just above its junction with the bulb, and the flask is swirled during the addition. An approximately equal volume of chloroform is added, and the flask is held in a steam bath and swirled until the solvent is brought just to a boil. The flask and its contents are cooled to room temperature, chloroform is added to the mark, and the contents are thoroughly mixed by repeated inversion of the flask and filtered through Schleicher and Schuell sharkskin paper which has previously been rendered lipid-free by exhaustive extraction with hot ethanol in a continuous extractor (10). A 20-ml. portion of the filtrate is pipetted at once for purification by one or the other of two procedures.

PURIFICATION PROCEDURE 1

This is a slightly modified application of the original method which Folch et al. (6) applied to brain lipid extracts. The 20-ml. aliquot is pipetted into a vial approximately 2 cm. in diameter and 8 cm. high, and water is added by means of a solvent pipet [Fig. 2 A, (9)] until the vial is full. Care is taken to avoid agitation of the chloroform-methanol solution by allowing the water to run slowly down the wall at the start, but some droplets of the solution are usually carried up and hang at the surface of the water. They may be dislodged by allowing water to drop on them after about half the water has been added and there is no further danger of agitating the solution. By means of a hemostat with wide, curved jaws the vial is lowered into a suitable rack at the bottom of a large (800 ml. or 1 l.) beaker which is nearly full of distilled water. Diffusion is allowed to proceed for at least 18 hr.

The vial is lifted from the beaker by means of the hemostat and clamped on the rod of the Lab Jack, and as much as possible of the upper phase is removed by means of a Van Slyke-Rieben pipet (11). The suction is previously adjusted so the rate of flow will just permit the counting of drops. The vial is raised at once until the tip of the pipet is about 1 to 2 mm. above the interface. Good illumination should be used, and a magnifier is desirable. The "fluff" which usually col-

176   W. M. SPERRY

lects in small amounts at the interface and sometimes on the wall
above it is easily loosened into suspension. If any tendency toward this
is seen, the vial is lowered a little. The vial is lowered quickly just as
the surface of the liquid reaches the tip. The agitation caused by suction at the surface is almost certain to cause suspension of the "fluff."
It is for this reason that the pipet tip is kept as close to the interface as
possible throughout the removal of the upper phase.

PURIFICATION PROCEDURE 2

This is a slightly modified application of the revised method of
Folch *et al.* (7, 8). The 20-ml. aliquot of the filtrate is pipetted into a
25-ml. glass-stoppered cylinder, previously cleaned with acid chromate
"cleaning mixture" and tested for tightness of the stopper [cf. (12), pp.
94, 95], 4 ml. of water are added, the stopper is moistened with water
and tightly inserted, and the contents are shaken vigorously for 1 minute. The cylinder is allowed to stand overnight or until the two phases
are clear except for some small droplets of the opposite phase which
usually adhere to the wall, particularly of upper phase below the
interface. These can sometimes, but not always, be freed by rapid
back and forth rotation of the cylinder while it is held in a vertical
position. Droplets of lower phase which may hang at the surface of
the upper phase can usually be dislodged by tapping the cylinder.
The upper phase is removed exactly as described in Procedure 1
except that somewhat less care is necessary since the absence of "fluff"
reduces the danger of loss.

Because the volume of the upper phase containing the nonlipid
impurities is small, in contrast to the conditions of Procedure 1, the
residue must be removed by washing. A solution for this purpose is
prepared as follows: 12 volumes of chloroform-methanol $(2 + 1)$ and
3 volumes of the $CaCl_2$ solution are thoroughly equilibrated in a
separatory funnel and allowed to stand until the phases have separated. The lower phase is discarded. The upper phase, which com-
prises about 40% of the total volume, and which has approximately
the same solvent concentration as the upper phase obtained in Procedure 2, is used as a wash solution; 3 ml. are added to the cylinder
with care to avoid mixing with lower phase. The pipet tip is held
against the wall just below the neck, and the solution is added slowly
as the cylinder is turned so the wall is washed. The cylinder is briskly
turned back and forth in a vertical position to mix the residue of the

original upper phase with the wash solution, which is removed as before. The washing procedure is repeated twice more.

TRANSFER AND EVAPORATION TO DRYNESS.

If Procedure 1 was used, 6 ml. of methanol are added to the vial and the mixture is stirred. A clear solution usually results, but a little more methanol may be required if an unusually large amount of upper phase was left. If Procedure 2 was used, 2 ml. of methanol are added to the cylinder. A clear solution results when the contents are mixed by shaking the cylinder. With either procedure, the purified lipid solution is transferred to a 50-ml. Erlenmeyer flask by means of apparatus C [Fig. 2 (9)]. (The external arm of tube S must be long enough to reach the bottom of the cylinder if Procedure 2 was used.) The solution is drawn over by gentle suction, controlled by a screw clamp on a side tube. The lower external wall of tube S is rinsed with a few drops of chloroform-methanol (2:1), and the wall of the vial or cylinder is rinsed with the same solvent, as follows: The tip of a solvent pipet [A, Fig 2 (9)] containing the solvent, is held against the wall of the vial just below the rim or the neck of the cylinder and solvent is expressed as the vial or cylinder is turned so a continuous film washes the entire wall. This requires very little solvent and only a few moments. The wash is sucked into the Erlenmeyer. The washing procedure is repeated four times more, five in all. The stopper is removed, the inner arm of the S tube is rinsed with a few drops of solvent, the stopper is reinserted, and a little more solvent is sucked over.

The flask containing the extract is clamped in a bath at about 40°C., and the solvent is removed with a stream of nitrogen admitted through apparatus B [Fig. 2 (9)]. If Procedure 1 was used, the time required will depend on the amount of aqueous upper phase which was not removed. The evaporation of the water may be facilitated by the addition of small portions of ethanol.

FILTRATION AND WEIGHING

As soon as the sample is dry, a small amount of chloroform-methanol is added as described above in the washing procedure. The solution is filtered by apparatus D [Fig. 2 (9)] into a 5-ml. volumetric flask which has previously been thoroughly cleaned, rinsed with a little chloroform-methanol (2 + 1), and dried and weighed as described below. The Erlenmeyer flask is washed five times as described above. The solvent is removed at 40°C. by a stream of nitrogen, the external surface of the flask is rinsed with distilled water, and the flask is thoroughly wiped with a clean towel and placed in a vacuum desiccator over Drierite. After evacuation, nitrogen is admitted to the desiccator, and it is evacuated again. This is repeated twice more, and the desiccator is left overnight in the balance room. The flask, usually with others, is placed in a Dri-Jar desiccator containing Drierite, allowed to equilibrate for about one-half hour, and weighed with the same sequence and timing used in obtaining the tare weight. One of the four flasks in a desiccator is kept as a control, and the weights of the other flasks are corrected for any change in its weight. This correction is usually small, but it may be appreciable if there is a marked change in atmospheric conditions.

Calculation

The weight of lipids in mg. multiplied by 125 gives the concentration in mg. per 100 ml.

NOTE: Checker S. L. J. studied the two methods described in his laboratory and found them to agree well with each other. The first procedure for purification was deemed more convenient, and was used extensively with the following minor changes made solely with the intent of providing even greater convenience in the clinical laboratory:

Specimen volume permitting, 2 ml. of serum was used per test with volumes of all solvents doubled. Whatman No. 1 filter paper was found to be as effective and rapid as the sharkskin paper for removing the protein precipitate from the extraction mixture. Furthermore, a blank determination carried through the entire procedure showed that it was not necessary to extract the Whatman paper before use. The filtrate in a 100-ml. beaker was immersed overnight in water as described. The supernatant aqueous phase was carefully removed with a cannula connected to a water aspirator by means of a rubber tube, thus eliminating the need for the Lab Jack. The chloroform phase was evaporated to dryness directly in the beaker with warming in a 40°C water bath and with the aid of nitrogen to facilitate evaporation and to minimize the possibility of oxidative degradation. It was unnecessary to add large amounts of methanol since there was no transfer from one container to another and a clear solution was therefore not required. This somewhat offset the delay in evaporating solvents which would be predicted in using a wide beaker rather than a narrow vial in the diffusion step thereby permitting not quite so complete removal of the supernatant aqueous phase. The residue, redissolved in chloroform-methanol was quantitatively filtered through a sintered glass funnel about 1 cm. in diameter fitted with a small piece of filter paper. An arrangement was devised to draw the solution into a weighed 10-ml. volumetric flask by suction. The filtrate was evaporated and the flask and residue dried in a vacuum desiccator as described with calcium chloride substituted for the Drierite. Use of the Dri-Jar desiccator was omitted.

The precision of the test with these changes was calculated to be ± 2.4% (95% confidence limits) from duplicates in a series of twenty-three experiments.

NOTE: Checker J. A. used a five ml. pipet to separate the phases and evaporated the chloroform-methanol phase in air. He found the method easy and reproducible in the errors of ± 1%.

Discussion

EXTRACTION

Serum is treated with methanol and chloroform is added to make the final concentration, 2 + 1, because a much more finely divided precipitate is thus obtained. It is quite possible that equally good extraction would be obtained if the serum were added to chloroform-methanol (2 + 1). The solvent is brought to a boil, not to improve the extraction, but to coagulate the proteins and improve the filtration. The same result could doubtless be obtained by allowing the mixture to stand at room temperature. Sharkskin paper has been used in our laboratory for many years for the filtration of lipid extracts. It is exceedingly fast, thus reducing the error due to evaporation, and it is sufficiently retentive to hold all of the protein if it is well coagulated.

PURIFICATION

The two procedures give the same result and the choice between them must be made on the basis of their advantages and disadvantages in terms of time and convenience. Procedure 1 requires more bench space to accommodate the large beakers, but considerably less working time since no washing is required. There is, however, greater danger of loss of lipids in the form of "fluff," and the drying requires a longer time because of the greater amount of water. With Procedure 2, the total time required for an analysis could be considerably shortened if centrifuging (cf. 8) instead of overnight standing were used to separate the phases, but the working time would be increased. Centrifuging would also avoid the retention of upper phase on the wall of the cylinder within the lower phase, but in practice no effect of this potential source of error has been observed.

In a previous description of this method (13), there were two deviations from the conditions recently prescribed by Folch et al. (8) for

Procedure 2. (*a*) If it is assumed that serum contains 93% water by volume, 4.07 instead of 4 ml. of water would be required under the conditions of this method to produce the specified 8:4:3 proportions of chloroform, methanol, and water, respectively, in the final mixture. (*b*) An aqueous solution containing 20 instead of 40 mg. of $CaCl_2$ per 100 ml. was used in preparing the upper phase used for washing. This resulted from a misunderstanding of a conversation with Dr. Folch. It was unlikely that the small difference in volumes of water would have an effect, but the use of too little $CaCl_2$ could well influence the results. Both points were tested in a series of duplicate experiments carried out on 20-ml. aliquots of an extract made from 10 ml. of pooled serum in a 250-ml. volumetric flask. The results with Procedure 1, the original (13) Procedure 2, and all possible combinations of the two variables were in close agreement. Also in one experiment the use of 4.00 ml. of 0.04% $CaCl_2$ instead of water at the first step in Procedure 2 gave a value in close agreement with the others. Despite these findings, which indicate that the original method gave correct results, the author has adopted 0.04% $CaCl_2$ on the basis of the much greater experience of Folch *et al.* with their procedure.

FILTRATION AND WEIGHING

If Procedure 1 is used a very small amount of precipitate is usually to be seen in the solution obtained after drying. This may be protein from proteolipids. No such precipitate is seen when Procedure 2 is used in conformity with the statement by Folch *et al.* (8) that proteolipids are not split on drying when their revised purification procedure is applied. This being true, filtration could be omitted if Procedure 2 is used, but this is not recommended. The filtration procedure is exactly the same as that for quantitative transfer with apparatus C [Fig. 2 (9)], and it takes no more time if the filter is properly prepared. It is well to use it, therefore, as a precautionary measure to remove dust and dirt which have a habit of being present despite all attempts to exclude them.

The procedure of admitting nitrogen to the desiccator was adopted to prevent oxidation as indicated by incomplete solubility in chloroform-methanol after weighing and by incomplete recovery of cholesterol by the Sperry-Webb method (14). It is probable that this step could be omitted if the desiccator were evacuated to low pressure by an oil pump instead of the water aspirator which we use.

The reliability of the results obtained with this method depend, of course, on the precision of weighing. The procedure will depend on the balance used and, perhaps, on the technique to which the analyst is accustomed. With the damped balance (Sartorius Selecta) which the author uses, it has been found desirable to time each weighing with a stop watch. The weight is recorded 1¼ minutes after the beam is released. A zero point reading is taken after each weighing at the same time interval after release of the beam, and a correction of half the change, if any, is applied to the weight. With these precautions, remarkably reproducible weights can be obtained.

It may not be necessary to leave the samples in the desiccator overnight. Recently, some preliminary weighings were made after only 1 to 2 hours in the desiccator. The flasks were then returned to the desiccator, which was evacuated and left overnight as prescribed in this method. The second weighings the next day agreed almost exactly with the first. This change in technique would have to be tested, however, before adoption in this method because the substances weighed were not serum lipids.

General

The method requires considerably more serum than do the oxidative methods, but there is no waste except for the small amount of filtrate which is discarded. If precautions are taken to prevent oxidation, the lipids are completely soluble in chloroform-methanol after weighing and may be used for any desired analyses of the lipid constituents. Since they are already in a volumetric flask, the solution may be made to volume, and aliquots may be taken.

REFERENCES

1. Bang, I., Verfahren zur titrimetrischen Mikrobestimmung der Lipoidstoffe. *Biochem. Z.* 91, 86–103; Die Mikrobestimmung der Blutlipoide. pp. 235–56 (1918).
2. Bloor, W. R., The determination of small amounts of lipide in blood plasma. *J. Biol. Chem.* 77, 53–73 (1928).
3. Bragdon, J. H., Colorimetric determination of blood lipides. *J. Biol. Chem.* 190, 513–7 (1951).
4. Christensen, H. N., The contaminants of blood phospholipides. *J. Biol. Chem.* 129, 531–7 (1939).
5. Folch, J., and Van Slyke, D. D., Nitrogenous contaminants in petroleum ether extracts of plasma lipides. *J. Biol. Chem.* 129, 539–46 (1939).
6. Folch, J., Ascoli, I., Lees, M., Meath, J. A., and LeBaron, F. N., Preparation of lipide extracts from brain tissue. *J. Biol. Chem.* 191, 833–41 (1951).

7. Folch, J., Lees, M., and Sloan-Stanley, G. H., A simple method for preparation of total pure lipide extracts from brain. *Federation Proc.* **13**, 209 (1954).
8. Folch, J., Lees, M., and Sloan-Stanley, G. H., A simple method for the isolation and purification of total lipides from animal tissues. *J. Biol. Chem.* **226**, 497–509 (1957).
9. Sperry, W. M., A method for the determination of total lipides and water in brain tissue. *J. Biol. Chem.* **209**, 377–86 (1954).
10. Sperry, W. M., Lipide excretion. III. Further studies of the quantitative relations of the fecal lipides. *J. Biol. Chem.* **68**, 357–83 (1926).
11. Van Slyke, D. D., and Rieben, W. K., Microdetermination of potassium by precipitation and titration of the phospho-12-tungstate. *J. Biol. Chem.* **156**, 743–63 (1944).
12. Sperry, W. M., Lipide analysis. In "Methods of Biochemical Analysis" (D. Glick, ed.). **2**, pp. 83–111. Interscience, New York, 1955.
13. Sperry, W. M., and Brand, F. C., The determination of total lipides in blood serum. *J. Biol. Chem.* **213**, 69–76 (1955).
14. Sperry, W. M., and Webb, M., A revision of the Schoenheimer-Sperry method for cholesterol determination. *J. Biol. Chem.* **187**, 97–110 (1950).

OXYGEN SATURATION OF BLOOD*

Submitted by: GEORGE W. JOHNSTON, Third U.S. Army Medical Laboratory, Fort McPherson, Georgia

Checked by: S. L. JACOBS, Bioscience Research Foundation, Los Angeles, California

Introduction

The rapid, accurate determination of the oxygen saturation of blood requiring only small samples has become of increasing importance in recent years in the clinical laboratory due to the introduction of cardiac catheterization as a routine procedure used in the diagnosis of heart defects and the development of surgical techniques for the correction of defects of the heart and great vessels.

Gordy and Drabkin in 1957 (1) described a simplified spectrophotometric procedure for the determination of the oxygen saturation of whole hemolyzed blood. In principle whole blood is hemolyzed by the addition of a strong hemolyzing agent such as saponin. The sample is transferred to an absorption cell of narrow light path (approximately 0.1 cm.), and light absorption measurements are made at two wavelengths. In the region of 650 mμ hemoglobin absorbs light to a much greater degree than does oxyhemoglobin, while at 805 mμ the two pigments absorb light equally (isosbestic). Assuming a two component system and absence of significant amounts of carboxyhemoglobin or methemoglobin, one may calculate the relative percentage of each of the two components in the mixture from light absorption measurements in these two regions.

The described method of sampling by use of tonometers is technically precise but somewhat inconvenient for clinical use, particularly where multiple sampling is desired.

In a note appended to the article of Gordy and Drabkin (1) Marsh described a simpler modification. He collected samples in oiled, heparinized syringes and added the hemolyzing agent, saponin, directly by means of a syringe and three-way metal stopcock. A simple, open, 1-cm. absorption cell fitted with a 9-mm. Lucite plug to reduce

---

* Based on the method of Gordy and Drabkin (1) and Johnston et al. (2).

the light path to 1 mm. was substituted for the more complicated anaerobic cell of Gordy and Drabkin.

In 1959, Johnston, Holtkamp, and Eve (2) reported a further modification in that a 1-ml. portion of the sample is transferred anaerobically to a 1-ml. tuberculin syringe by means of a three-way metal stopcock. Hemolyzing and reducing solutions may then be added through the same stopcock and mixed with the sample before transfer to an absorption cell. This technique permits the rapid handling of multiple samples and provides aliquots for duplicate analyses or other analytical measurements such as hemoglobin or hematocrit estimations.

Reagents

1. Hemolyzing solution (solution A). Dissolve 6.0 g. of white saponin in 20 ml. of 0.10% sodium carbonate solution. Centrifuge to clear of sediment.

NOTE: Each brand and lot number of saponin should be pretested to insure complete hemolysis of blood under the conditions of the procedure. Most lots of the Coleman and Bell product gave satisfactory results.

NOTE: Checker S. L. J. used 15% Triton X-100 (Rohm and Haas Co., Phila., Pa.) in 0.1% Na_2CO_3 successfully as a substitute for saponin and found it more uniform from batch to batch than saponin. Since this reagent gave negligible readings at 650 and 805μ it was replaced by water for the spectrophotometric reference.

2. Hemolyzing and reducing solution (solution B). Transfer 0.50 g. of sodium hydrosulfite ($Na_2S_2O_4$) to a 10-ml. tube fitted with a tight fitting rubber or glass stopper. Add 5.0 ml. of solution A, stopper quickly, and mix until solution is complete. Centrifuge to clear of sediment.

3. Heparin, sodium, injection, 1000 units per ml.

4. Petrolatum, white, U.S.P.

5. Mercury, U.S.P., clean.

Apparatus

1. Syringes, 5 ml.
2. Syringes, 1 ml. tuberculin.
3. Stopcocks, metal, Ayer three-way, to fit syringes.
4. Needles, hypodermic, No. 22, 3 in.
5. Spectrophotometer capable of operating in the region of 600–825

mμ with high resolution. The Beckman DU and Bausch and Lomb Spectronic 20 instruments have been found satisfactory.

6. Absorption cells, 1 cm. for the above instrument.

7. Cell spacers 9 mm. custom fitted to 1-cm. absorption cells.

NOTE: The 9-mm. silica cell spacers manufactured by Beckman Instruments Corporation, Fullerton, California and Pyrocell, New York, and custom made Lucite cell spacers have been found satisfactory.

Procedure

Blood samples are collected in sterile, heparinized 5-ml. syringes, the dead space of the needle and syringe being filled with heparin solution. To facilitate mixing, mercury is added to the sample in the following manner (see Fig. 1): The needle is removed and the mer-

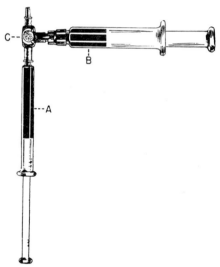

FIG. 1. Mixing Syringe (A) attached to sample syringe (B) by means of three-way metal stopcock (C).

cury is injected from a syringe and needle through the nozzle of the sample syringe. The needle is replaced, air bubbles are ejected, and the sample is sealed from the air by impaling a soft rubber stopper on the needle. Samples are immersed in an ice bath until analysis, which should be performed no later than a few hours after collection.

One-milliliter tuberculin syringes are prepared for use by first lubricating the plungers with white petrolatum. Mercury is forced in and out of the syringe to remove air bubbles, and the syringe is one-third filled with mercury. The syringe is held in a vertical position, nozzle uppermost, an Ayer three-way metal stopcock equipped with a bent needle is attached and the excess mercury is forced from the assembly. The stopcock is turned through 90 degrees leaving the dead space of the assembly filled with mercury.

The syringe containing the well-mixed sample is attached to the side arm and air bubbles are ejected through the stopcock. The stopcock is turned through 90 degrees and blood is added to the 1.0-ml. mark of the mixing syringe. A 0.20-ml. quantity of hemolyzing solution (solution A) is then added from a tuberculin syringe through the stopcock. The mixture is shaken end-to-end for 2–4 minutes. The hemolyzed sample is transferred by means of a 3-inch No. 22 needle bent at a right angle to a 1-cm. absorption cell which has been fitted with a close fitting 9-mm. cell spacer to reduce the light path to 1 mm. Absorbance readings are made at 650 and 805 mμ using solution A in a similar plugged cell as a reference.

Similar readings are made on a sample completely saturated with oxygen and on another sample after complete reduction of the HbO_2 to Hb by substituting the hemolyzing and reducing solution (solution B) for solution A.

Calculation

$$\text{Fraction of } Hbo_2 = \frac{\dfrac{D_{650}Hb}{D_{805}Hb} - \dfrac{D_{650}}{D_{805}}}{\dfrac{D_{650}Hb}{D_{805}Hb} - \dfrac{D_{650}HbO_2}{D_{805}HbO_2}}$$

Discussion

The ratios

$$\frac{D_{650}Hb}{D_{805}Hb} \text{ and } \frac{D_{650}HbO_2}{D_{805}HbO_2}$$

are relatively independent of hemoglobin concentration and are constant for a given set of experimental conditions. In a series of seventeen analyses Johnston et al. (2) found the following values for the ratios using a Beckman DU spectrophotometer equipped with 1-cm. absorption cells plugged with 9-mm. silica cell spacers:

	Mean	Standard deviation
Hb $\dfrac{D_{650}}{D_{805}}$	3.97	0.06
HbO$_2$ $\dfrac{D_{650}}{D_{805}}$	0.50	0.06

NOTE: Checker S. L. J. found Hb $\dfrac{D_{650}}{D_{805}}$ to be 4.01 ± 0.145 and HbO$_2$

$\dfrac{D_{650}}{D_{805}}$ to be 0.58 ± 0.067 for 20 determinations.

In twelve analyses of blood for oxygen saturation a mean difference of 3% was observed between the spectrometric and gasometric results with the spectrometric being in all cases equal to or greater than the latter.

ADDENDUM: G. W. J. in answer to a question wrote the following which is added to this method because it could be useful to others:

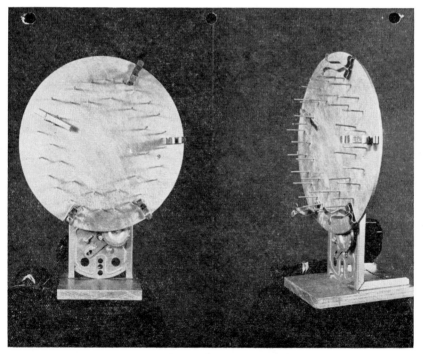

FIG. 2. Homemade mechanical rotator.

It was our practice to experimentally verify these values at least once each working day using a single sample of blood as follows:

1. Transfer 3 to 5 ml. of a well mixed blood sample to a 50 ml. bottle or flask. Replace the air with oxygen from a tank. Stopper and rotate the container for 10 minutes or so, making additional oxygen replacements at intervals.

We used a homemade mechanical rotator for all mixing operations which was made from a phonograph turntable motor.

2. Transfer 1 ml. portions of the well mixed sample to each of two previously prepared tuberculin syringes.

3. Add 0.2 ml. of solution A to one, mix and make optical density measurements at 650 and 805 mμ as above. This is the 100 per cent oxygenated sample.

4. Add 0.2 ml. of solution B (containing hydrosulfite) to the other, mix and make optical density measurements at 650 and 805 mμ. This is the zero per cent oxygen saturation sample.

Calculations are made as follows:

$$\text{fraction } HbO_2 = \frac{\dfrac{D_{650}Hb}{D_{805}Hb} - \dfrac{D_{650} \text{ Unknown}}{D_{805} \text{ Unknown}}}{\dfrac{D_{650}Hb}{D_{805}Hb} - \dfrac{D_{650}HbO_2}{D_{805}HbO_2}}$$

EXAMPLE:

Reduced sample

$D_{650}Hb = 0.487$

$D_{805}Hb = 0.126$

Oxygenated sample

$D_{650}HbO_2 = 0.057$

$D_{805}HbO_2 = 0.128$

Unknown sample

D_{650} Unknown $= 0.237$

D_{805} Unknown $= 0.127$

$$\text{fraction } HbO_2 = \frac{\dfrac{0.487}{0.126} - \dfrac{0.237}{0.127}}{\dfrac{0.487}{0.126} - \dfrac{0.057}{0.128}}$$

$$= \frac{3.87 - 1.87}{3.87 - 0.44}$$

$$= \frac{2.00}{3.43}$$

$$= 0.58 \text{ or } 58 \text{ per cent saturation}$$

A page from the notebook looks like this:

Sample

1   $\dfrac{0.237}{0.127} = 1.87$ $\begin{array}{r} 3.87 \\ 1.87 \\ \hline 2.00 \end{array}$ $\dfrac{2.00}{3.43} = 0.58 \text{ or } 58\%$

Hb $\dfrac{0.487}{0.126} = 3.87$ $\begin{array}{r} 3.87 \\ 0.44 \\ \hline 3.43 \end{array}$

HbO$_2$ $\dfrac{0.057}{0.128} = 0.44$

REFERENCES

1. Gordy, E., and Drabkin, D. L., Spectrophotometric studies XVI: Determination of the oxygen saturation of blood by a simplified technique, applicable to standard equipment. *J. Biol. Chem.* **227**, 285–299, (1957).
2. Johnston, G. W., Holtkamp, F., and Eve, J. R., The spectrophotometric determination of the oxygen saturation of blood. *Clin. Chem.* **5**, 421–425 (1959).

DETERMINATION OF SERUM INORGANIC PHOSPHORUS*

Submitted by: R. L. Dryer and J. I. Routh, Department of Biochemistry, University of Iowa College of Medicine, Iowa City, Iowa

Checked by: Ruth D. McNair, Providence Hospital, Detroit, Michigan

Introduction

Methods for the colorimetric determination of phosphorus depend almost entirely on the reduction of phosphomolybdic acid to form a blue pigment which is presumably a mixture of molybdenum oxides. Variations in methodology revolve about the choice of a reducing agent and, to a lesser degree, on the use of solvents to concentrate the pigment formed. The method of Fiske and SubbaRow (1) in which 1,2,4-aminonaphtholsulfonic acid (ANS) is utilized has probably received the widest acceptance. Most sources of ANS furnish material which must be carefully purified by a tedious process of recrystallization before it is suitable for use. Even after purification, the shelf-life of ANS is limited by the ease with which it undergoes air oxidation. Furthermore, the rate of reduction of phosphomolybdate by ANS is slow, and its use as a reducing agent requires at least 45 minutes for the development of maximal absorbances.

Substitution of stannous chloride for ANS has been suggested by Kuttner and Cohen (2) and by Shinowara et al. (3). This reagent gives somewhat greater absorbances than ANS, and the reduction is effected more rapidly, but the absorbance of the solutions never become linear with respect to time. Stannous chloride is a very vigorous reducing agent, and unless great care is taken will reduce silicomolybdic acid as well as phosphomolybdic acid. Unless the pH is carefully controlled serious errors may be introduced by employing stannous chloride, especially if the molybdate solutions have been stored in glass. Beveridge and Johnson (4) used a mixed reagent which included molybdate and hydrazine sulfate, and Taussky and Shore (5) employed ferrous sulfate. Both of these methods require that the reducing reagent be freshly prepared. Gomori (6) suggested and

* Based on the method of Dryer et al. (8).

191

192

Power (7) has recommended the use of p-methylaminophenol sulfate (Elon). Elon is considerably more stable than ANS and is readily available in a state of adequate purity. Its rate of reduction is comparable to that of ANS, requiring from 45 minutes to 1 hour for production of maximal absorbances.

Semidine hydrochloride (N-phenyl-p-phenylenediamine hydrochloride) is suitable for the reduction of phosphomolybdic acid and offers certain advantages. It is, like Elon, available in a suitable state of purity, it produces absorbances equal to or greater than those given by ANS, and the maximal absorbances are produced in 10 minutes (8). By virtue of its oxidation-reduction potential, there is a minimal likelihood that silicomolybdic acid will be reduced by this reagent.

Reagents

1. *Trichloroacetic acid, 10%.* Dissolve 100.0 g. of crystalline trichloroacetic acid (A.R.) in water and dilute to 1 l.

2. *Ammonium molybdate, 0.008 M.* Dissolve 9.887 g. of ammonium molybdate [$(NH_4)_6Mo_7O_{24} \cdot 4H_2O$] (A.R.) in approximately 750 ml. of water. If an appreciable residue persists, add a few drops of conc. sulfuric acid to effect complete solution.

NOTE: Not more than 0.25 ml. of acid should be required for the preparation of the solution just described. If the solid cannot be dissolved by addition of the stated amount of acid, the ammonium molybdate may be regarded as unsuitable for this purpose. Do not use sulfuric acid unless it is necessary.

Dilute the solution to 1 l. and store in a polyethylene bottle.

3. *Semidine hydrochloride, 50 mg. per 100 ml., in 1% sodium bisulfite.* Weigh out 250.0 mg. of solid semidine hydrochloride and transfer the solid to a 500-ml. volumetric flask. Add approximately 0.5 ml. of methanol or ethanol in such a way that the solvent wets the powder as completely as possible. Dissolve the wetted powder by the addition, with shaking, of a 1% solution of sodium bisulfite. This is prepared by solution of 6.0 g. of sodium bisulfite (A.R.) in 600 ml. of water.

NOTE: Semidine hydrochloride is available from Distillation Products Industries, Rochester, New York, as N-phenyl-p-phenylenediamine, No. 2043. This solid salt is wet by aqueous solutions with some difficulty, which can be avoided by the treatment with methanol or ethanol as outlined above. Once the solid is wet it should dissolve in the bisulfite with ease. An occasional sample may leave a slight trace of undissolved residue. This may be filtered off through either glass wool or any acid-washed filter paper. Air oxidation is minimized by the use of sodium bisulfite as the solvent, but the semidine solution is still slowly oxidized by molecular oxygen. Make fresh solution every 3 to 4 weeks, or as soon as it takes on a yellow color. Since the oxidation is photocatalyzed, avoid excessive exposure to light.

4. Stock phosphorus standard (100.0 mg. P per 100 ml.). Dissolve 438.1 mg. of potassium dihydrogen phosphate (A.R.) in 100.0 ml. of 10% trichloroacetic acid.

5. Working phosphorus standard (1.0 mg. P per 100 ml.). Dilute 1 ml. of the stock standard to 100.0 ml. with 10% trichloroacetic acid. The working standard corresponds to a serum concentration of 10 mg. per 100 ml. of inorganic phosphorus when employed as outlined in the procedure.

Procedure

Place 1.80 ml. of 10% trichloroacetic acid in a 12 or 15 ml. centrifuge tube. Add 0.20 ml. of serum, mix well, and allow to stand for at least 5 minutes. Centrifuge for 5 minutes at 1500 r.p.m. or until the supernatant is clear and free of particles floating on the surface. Transfer 1.0 ml. of the supernatant to a test tube cuvette.

Prepare a reagent blank by placing 1.0 ml. of 10% trichloroacetic acid in a cuvette. At the same time prepare a suitable standard by placing 1.0 ml. of the working standard solution in another cuvette. To all of the cuvettes add 0.20 ml. of the molybdate reagent, and mix well. Finally add to each cuvette 2.0 ml. of the semidine reagent and mix well by shaking. Allow the cuvettes to stand for 10 minutes, then measure the absorbance of each solution at a wavelength of 700 mμ or 345 mμ (see Discussion), setting the zero of the absorbance scale with the reagent blank.

Calculation

The absorbances given by solutions containing up to 15 mg. per 100 ml. obey the Beer-Lambert Law, so the conversion of absorbance to phosphorus concentrations may be achieved by simple application of the formula

$$C_u = C_s \frac{A_u}{A_s}$$

where C_u is the inorganic phosphorus content of the serum sample, C_s is the serum equivalent concentration of the working standard, A_u is the absorbance (optical density) of the unknown serum sample, and A_s is the absorbance of the standard solution.

A series of standard solutions equivalent to serum concentrations from 1 to 9 mg. per 100 ml. can be prepared by diluting aliquots of from 1 to 9 ml. of the working phosphorus standard to a volume of 10 ml. with 10% trichloroacetic acid. Transfer a 1.0-ml. aliquot of each standard dilution to a cuvette and treat as above with the molybdate and semidine reagents.

A standard equivalent to 5 mg. P per 100 ml. of serum will give an absorbance of 0.159 at 700 mμ in a Beckman DU spectrophotometer, employing a slit width of 0.05 and using a 1 cm. cuvette. At 345 mμ the absorbance of the same solution will be 0.412. The same solution will give an absorbance of 0.18 at 700 mμ in a Coleman Jr. spectrophotometer, using 12-mm. cuvettes. Other instruments will give values which approximate these, depending on the individual instrument parameters.

Discussion

Most methods for the colorimetric determination of phosphorus measure absorbance or transmittance at or near the red end of the visible spectrum, usually at some point between 650 and 750 mμ. Since the spectral curve of the blue solutions is quite flat in this area, the exact wavelength is not critical. We have observed that between 340 and 400 mμ the absorbance of the blue solution is several fold greater than at any point in the visible spectrum. The curious fact is that this difference is due at least in part to the mixed molybdenum oxides themselves, since the increased absorbance can be seen with a number of reducing agents which were examined. Details have been reported elsewhere (8). We recommend the use of the 345 mμ wavelength where maximal sensitivity is required. Since the spectral curve is quite steep in this area the narrowest possible slit setting should be employed.

The absorbances of the colored solutions remain constant for at least 1.5 hours, which is as long as we have observed them.

Two modifications of the above procedure have been developed, one for the determination of serum phosphatase and one for the determination of lipid phosphorus (8).

Results

The range of serum inorganic phosphorus levels obtained by this method are substantially in accord with values obtained by other common procedures. A series of 60 analyses made on apparently healthy medical students in the fasting state gave a range of 2.6–4.8 mg. P per 100 ml.

It must be emphasized that reliable values of serum inorganic phosphate can only be obtained on properly collected samples. To avoid the effects of glycolysis in shed blood the serum should be separated from the cells as promptly as possible.

REFERENCES

1. Fiske, C. H., and SubbaRow, Y., The colorimetric determination of phosphorus. *J. Biol. Chem.* **66**, 375–400 (1925).
2. Kuttner, T., and Cohen, H. R., The micro estimation of phosphate and calcium in pus, plasma, and spinal fluid. *J. Biol. Chem.* **75**, 517–531 (1927).
3. Shinowara, G. Y., Jones, L. M., and Reinhart, H. L., The estimation of serum inorganic phosphate and "acid" and "alkaline" phosphatase activity. *J. Biol. Chem.* **142**, 921–933 (1942).
4. Beveridge, J. M. R., and Johnson, S. E., The determination of phospholipid phosphorus. *Can. J. Res. Sect. E*, **27**, 159–63 (1949).
5. Taussky, H. H., and Shore, E., A microcolorimetric method for the determination of inorganic phosphorus. *J. Biol. Chem.* **202**, 675–685 (1953).
6. Gomori, G., A modification of the colorimetric phosphorus determination for use with the photoelectric colorimeter. *J. Lab. Clin. Med.* **27**, 955–960 (1942).
7. Power, M. H., Inorganic phosphate. Standard methods of clinical chemistry **1**, 84–87 (1953).
8. Dryer, R. L., Tammes, A. R., and Routh, J. I., The determination of phosphorus and phosphatase with N-phenyl-p-phenylenediamine. *J. Biol. Chem.* **225**, 177–183 (1957).

THE ESTIMATION OF SEROTONIN IN BIOLOGICAL MATERIAL*

Submitted by: HERBERT WEISSBACH, Laboratory of Clinical Biochemistry, National Heart Institute, National Institutes of Health, Public Health Service, U.S. Department of Health, Education and Welfare, Bethesda, Maryland

Checked by: EDNA ANDREWS, Metropolitan Hospital, Detroit, Michigan
              HARRIET T. SELIGSON, Yale University, New Haven, Connecticut

Although serotonin (5-hydroxytryptamine; 5-HT) has only recently been isolated and characterized (1) much information has been obtained concerning both its biosynthesis and metabolism (2). The exact role that this indoleamine has in the body has not been ascertained although it is known that it is widely distributed in animal tissues such as the intestine, brain, blood platelets, lungs, and skin of certain species. The early assays for this compound relied on bioassay techniques (3, 4, 5, 6) which are still employed in some laboratories. The chemical methods described in this report have been found to be most satisfactory in this laboratory and are also widely used in other laboratories.

Principle

The unique fluorescence characteristics of 5-HT (7) have made it possible to determine it directly in many tissue extracts after precipitation of the proteins with zinc hydroxide (8). This method is applicable to most tissues containing more than 1 μg. of 5-HT per gram and to blood containing over 1 μg. per ml. Although it is the method of choice for its simplicity, it lacks specificity since all 5-hydroxyindoles have the same fluorescence characteristics. However, 5-HT is usually the only hydroxyindole present in tissues. When the amounts of 5-HT present are small, and interfering materials are present, it is necessary to purify and concentrate the tissue extracts before assaying 5-HT. This is best done by extraction of the 5-HT into n-butanol from a salt-saturated alkaline solution. The 5-HT can then be re-extracted

* Based on the method of Udenfriend et al. (9).

197

from the butanol into a small volume of dilute HCl (9, 10). The 5-HT in the final acid is then assayed fluorometrically.

Until recently it was not possible to measure the fluorescence of compounds, such as serotonin, that are activated in the far ultraviolet region. The development of the spectrophotofluorometer (11) has made it possible to extend fluorescent measurements into this portion of the ultraviolet region, since this instrument is capable of delivering high intensity monochromatic light for activation at all wavelengths from approximately 240–800 mμ, and can carry out automatic spectral analysis of the resulting fluorescence throughout this same range. Spectrophotofluorometers are now available commercially from the American Instrument Co., Silver Springs, Md., and the Farrand Optical Co., Yonkers, New York.

Activation of 5-HT occurs maximally at 295 mμ,[1] and in dilute acid or at neutral pH the fluorescence maximum is at 350 mμ. However, in 3 N HCl the fluorescence shifts and a new peak appears at 550 mμ. This shift of fluorescence in strong acid is a function of the phenolic group and distinguishes 5-hydroxyindoles from other closely related compounds. In solution it is possible to measure as little as 0.05 μg. of 5-HT with fluorescence techniques.

When large amounts of 5-HT are present in the tissue (more than 20 μg. per g.), as in animal experiments in which 5-HT or its precursor, 5-hydroxytryptophan, have been administered, or in malignant carcinoid tumors (12), it is possible to measure the 5-HT colorimetrically utilizing the reaction with 1-nitroso-2-naphthol (9). This is usually performed following the butanol extraction procedure.

The blood levels of 5-HT in man and many other animals are too low to be assayed by the direct protein precipitation method, and poor recoveries are obtained if one attempts to extract 5-HT from whole blood into butanol. However, the bulk of the 5-HT in the blood is present in the blood platelets and it is possible to isolate these blood cells and assay them for 5-HT. It has been found most convenient to correlate the amount of 5-HT present to the protein content of these cells (μg. of 5-HT per mg. of platelet protein).

Reagents and Equipment

1. Protein precipitation: 10% zinc sulfate H_2O, and 1 N NaOH.

[1] All wavelength data are presented as uncorrected instrumental readings. Wavelengths of maximal activation and fluroescence should be determined on each instrument by a comparison with appropriate standards.

2. *Extraction procedure.*

(a) *Borate buffer.* To 94.2 g. of boric acid dissolved in 3 l. of water add 165 ml of 10 N NaOH. The buffer solution is then saturated with purified n-butanol and NaCl by adding these substances in excess and shaking. Excess n-butanol is removed by aspiration and excess salt is allowed to settle. The final pH should be about 10.0.

(b) *n-Butanol.* Reagent grade n-butanol is purified by first shaking with an equal volume of 0.1 N NaOH, then with an equal volume of 0.1 N HCl, and finally twice with distilled water.

(c) *Heptane.* Practical grade heptane is treated in the same manner as the n-butanol.

3. *Colorimetric assay.*

(a) *1-Nitroso-2-naphthol reagent.* Dissolve 0.1% 1-nitroso-2-naphthol in 95% ethanol.

(b) *Nitrous acid reagent.* To 5 ml. of 2 N N_2SO_4 is added 0.2 ml. of 2.5% $NaNO_2$. This reagent should be prepared fresh daily.

(c) *Ethylene dichloride.*

4. *Fluorescence measurements.* Aminco-Bowman or Farrand spectrophotofluorometer. It should be possible to assay for 5-HT with a filter instrument capable of exciting at 280–300 mμ and of detecting fluorescence in the 530–550 mμ region with a sensitive instrument and suitable filter.

5. *5-HT stock solution.* Serotonin creatinine sulfate is commercially available. The indoleamine is 43.6% of the complex. Solutions of 100 μg. per ml. of 5-HT are stable for several weeks when kept in the cold, but more dilute solutions should be freshly prepared.

Procedures

1. DIRECT METHOD FOR 5-HT IN TISSUES AND BLOOD (8)

In assaying for 5-HT in blood 1 ml. is collected in a siliconized tube containing anticoagulant and is hemolyzed by the addition of 7 ml. of water. Tissues are homogenized in at least 2 volumes of 0.1 N HCl and an aliquot, corresponding to no more than 2 g., is diluted to 8 ml. with water. The proteins are then precipitated by the addition of 1 ml. of 10% zinc sulfate followed by 0.5 ml. of 1 N NaOH. After 5 minutes the tubes are shaken and centrifuged at 2500 r.p.m. for 20 minutes. One ml. of the clear supernatant fluid is transferred to a quartz cuvette containing 0.3 ml. of 12 N HCl. The fluorescence is measured in the spectrophotofluorometer (activation wavelength, 295 mμ; fluorescent wavelength, 550 mμ). Known amounts of serotonin and a reagent blank in which the tissue aliquot is replaced by water are carried through the entire procedure. When serotonin is added to tissue homogenates, it is recovered to the extent of 85–100%.

2. Extraction Procedure for 5-HT (9, 10)

One part of tissue is homogenized in 2 parts of 0.1 N HCl and an aliquot containing 0.2–5.0 μg. of 5-HT is transferred to a 60-ml. glass-stoppered bottle containing 3 ml. of borate buffer pH 10.0. If necessary adjust the pH to between 9 and 10 by the addition of anhydrous sodium carbonate. It is not crucial if the pH is slightly below 10 but it is important that it does not exceed 10.5. The solution is diluted to 10 ml. and 3–4 g. of NaCl are added to saturate the aqueous phase with salt. Fifteen milliliters of butanol are added and the bottle shaken for 10 minutes. After centrifugation the fluid is decanted from the solid material into another bottle and the aqueous phase is removed by aspiration. The butanol is then washed with an equal volume of borate buffer pH 10 to remove any neutral or acidic 5-hydroxyindoles such as 5-hydroxytryptophan. Ten milliliters of the butanol phase are then transferred to a 40-ml. glass-stoppered shaking tube containing 15 ml. of heptane and 2.5 ml. of 0.1 N HCl. The tube is shaken for 5 minutes, centrifuged, and 1 ml. of the dilute acid is transferred to a test tube containing 0.3 ml. of 12 N HCl. The solution is assayed fluorometrically in the spectrophotofluorometer; activation, 295 mμ; fluorescence, 550 mμ. Standards, recoveries, and a reagent blank are treated in the same way. 5-HT added to tissues is recovered to the extent of 92–106%.

When large amounts of 5-HT are present (20–150 μg.) the extraction procedure (to insure specificity) can be followed by colorimetric assay. Two milliliters of the final acid are transferred to a 12-ml. glass-stoppered centrifuge tube containing 1 ml. of the nitroso-naphthol reagent. After shaking, 1 ml. of nitrite reagent is added and the tube is shaken again and allowed to stand for 10 minutes at room temperature. Following this, unreacted nitroso-naphthol is removed by shaking with 5 ml. of ethylene dichloride. After centrifugation at low speed the aqueous phase is transferred to a cuvette and the optical density measured at 540 mμ. The color is stable for hours, and the optical density is proportional to 5-HT concentration up to 150 μg.

3. 5-HT Determination in Blood Platelets

In most species the 5-HT levels in blood are too small to be assayed by the direct method. The following procedure for the isolation of platelets and determination of their 5-HT content is recommended. About 10 ml. of blood are collected in siliconized glassware using 0.1 volume of 1% versene–0.7% saline as anticoagulant. All subsequent steps in the isolation of platelets (carried out at 3°C. using siliconized glassware) are performed according to the procedure of Dillard *et al.* (13). The blood is centrifuged at 600 r.p.m. in an International No. 1 centrifuge for 45 minutes. The platelet rich plasma is transferred to another centrifuge tube and centrifuged at 2000 r.p.m. for 30 minutes. The plasma is decanted and the sedimented platelets are carefully suspended in 5 ml. of cold saline to wash out adhering blood proteins. The tube is centrifuged at 2500 r.p.m. for 20 minutes and the saline discarded. This washing procedure is repeated once more and then the platelets are suspended in about 3.5 ml. of 0.02 N HCl. Gentle agitation causes lysing of the platelets. One ml. is then removed for 5-HT assay. It is diluted to 2 ml. with water and the proteins are precipitated by the addition of 0.2 ml. of 10% zinc sulfate and 0.1 ml. 1 N NaOH. After centrifugation the supernatant fluid is assayed fluorometrically; activation, 295 mμ; fluorescence, 350 mμ. Standards and a reagent blank are treated in the same manner. A second aliquot of the lysed platelet solution is assayed for protein according to the method of Sutherland *et al.* (14). Results are expressed as μg. of 5-HT per mg. of platelet protein.

NOTE: Protein may be measured by the procedure described in the volume for cerebrospinal fluid protein.

Discussion and Results

The direct method for the assay of 5-HT is applicable to most tissues containing more than 1 μg. of 5-HT per g. of tissue. Thus, it can be used to measure 5-HT in the intestines of most species, in the blood of rabbits, chickens, and patients with malignant carcinoid, and in the lungs of certain species. The extraction procedure, although more time consuming, insures specificity[2] and can be used for all tissues except blood. For the blood of those species containing less than 1 μg. of 5-HT per ml. the platelet assay is recommended, although Waalkes (15), in a recent publication, has been able to assay 5-HT in human

[2] The N-methyl derivatives of 5-HT have extraction and fluorescence characteristics similar to those of 5-HT, but have not been shown to be present in animal tissues.

202

                     H. WEISSBACH

blood by combining the zinc hydroxide precipitation and extraction procedures.
The 5-HT levels in various tissues are summarized in Tables I and II.

TABLE I
5-HT CONTENT OF VARIOUS TISSUES AND THE RECOMMENDED PROCEDURE[a]

Tissue	5-HT content (μg. per g. or ml.)	Method employed
Rat lung	3	Direct or extraction
Rabbit intestine	10	Direct or extraction
Guinea pig intestine	6	Direct or extraction
Rat intestine	1.5	Direct or extraction
Mouse intestine	5.3	Direct or extraction
Guinea pig lung	0.2	Extraction
Rabbit blood	4–6	Direct
Chicken blood	1–3	Direct
Carcinoid blood	1–3	Direct
Carcinoid tumor	1000	Direct or extraction

[a] From Udenfriend et al. (16).

TABLE II
5-HT IN PLATELETS[a]

Species	(μg. 5-HT per mg.) Platelet protein
Guinea pig	0.4
Rabbit	7–9
Dog	0.6–1.5
Human[b]	0.1–0.4[b]

[a] Taken from Udenfriend, S. et al. (16).
[b] Equivalent to 0.05–0.2 μg. of 5-HT per ml. of whole blood.

REFERENCES

1. Rapport, M. M., Serum vasoconstrictor (Serotonin). V. The presence of creatinine in the complex. A proposed structure of the vasoconstrictor principle. J. Biol. Chem. 180, 961–969 (1949).
2. Udenfriend, S., Titus, E., Weissbach, H., and Peterson, R. E., Biogenesis and metabolism of 5-hydroxyindole compounds. J. Biol. Chem. 219, 335–344 (1956).
3. Page, I. H., and Green, A. A., II. Perfusion of rabbit's ear for study of vasoconstrictor substances. Methods Med. Res. 1, 123–129 (1948).
4. Dalgleish, C. E., Toh, C. C., and Work, T. S., Fractionation of the smooth muscle stimulants present in extracts of gastro-intestinal tract. Identification

of 5-hydroxytryptamine and its distinction from substance P. *J. Physiol.* (*London*) **120**, 298–310 (1953).

5. Erspamer, V., Pharmacological studies on enteramine (5-hydroxytryptamine). IX. Influence of sympathomimetic and sympatholytic drugs on the physiological and pharmacological actions of enteramine. *Arch. Intern. Pharmacodyn.* **93**, 293–317 (1953).

6. Twarog, B. M., and Page, I. H., Serotonin content of some mammalian tissues and urine and a method for its determination. *Am. J. Physiol.* **175**, 157–161 (1953).

7. Udenfriend, S., Bogdanski, D. F., and Weissbach, H., Flourescence characteristics of 5-hydroxytryptamine (Serotonin). *Science* **122**, 972–973 (1955).

8. Weissbach, H., Waalkes, T. P., and Udenfriend, S., A simplified method for measuring Serotonin in tissues; Simultaneous assay of both Serotonin and Histamine. *J. Biol. Chem.* **230**, 865–871 (1958).

9. Udenfriend, S., Weissbach, H., and Clark, C. T., The estimation of 5-hydroxytryptamine (Serotonin) in biological tissues. *J. Biol. Chem.* **215**, 337–344 (1955).

10. Bogdanski, D. F., Pletscher, A., Brodie, B. B., and Udenfriend, S., Identification and assay of Serotonin in brain. *J. Pharmacol. Exptl. Therap.* **117**, 82–88 (1956).

11. Bowman, R. L., Caulfield, P. A., and Udenfriend, S., Spectrophotofluorometric assay in the visible and ultraviolet. *Science* **122**, 32–33 (1955).

12. Sjoerdsam, A., Weissbach, H., and Udenfriend, S., A clinical, physiological and biochemical study of patients with malignant carcinoid (Argentaffinoma). *Am. J. Med.* **20**, 520–531 (1956).

13. Dillard, G. H. L., Brecher, G., and Conkite, E. P., Separation, concentration and transfusion of platelets. *Proc. Coc. Exptl. Biol. Med.* **78**, 796–799 (1951).

14. Sutherland, E. W., Cori, C. F., Haynes, R., and Olsen, N. S., Purification of the hyperglycemic-glycogenolytic Factor from insulin and from gastric mucosa. *J. Biol. Chem.* **180**, 825–837 (1949).

15. Waalkes, T. P. Determination of Serotonin (5-hydroxytryptamine) in human blood. *J. Lab. Clin. Med.* **53**, 824–829 (1959).

16. Udenfriend, S., Weissbach, H., and Brodie, B. B., Assay of Serotonin and Related Metabolities, Enzymes, and Drugs *In* "Methods of Biochemical Analysis," Vol. 6, pp. 95–130. (D. Glick, ed.). Wiley (Interscience), New York (1958).

SOME APPLICATIONS OF STATISTICS TO CLINICAL CHEMISTRY

Submitted by: RICHARD J. HENRY, Director, Bio-Science Laboratories, 12330 Santa
Monica Boulevard, Los Angeles 25, California
ROBERT L. DRYER, Department of Biochemistry, State University of
Iowa, Iowa City, Iowa

Introduction

A continuous variable is a measurement or observation which, within practical limits, can take any value. Clinical chemists are concerned with continuous variables principally in two ways: (1) variation in the level of the substance measured, e.g., blood glucose level. Delineation of the normal range lies in this category; (2) error of measurement. A complete chemical analysis, which is nothing more than a system of manipulations and measurements has associated with it a corresponding error, the magnitude of which depends on the law governing the combination of the individual errors. The individual errors consist of effects of technique, chemicals, instruments, and certain less easily defined factors inherent in the nature of measurements. Good technique and equipment may decrease the over-all error, but never completely eliminate it.

Such problems cannot be approached quantitatively without the use of some simple statistical concepts and tools. The purpose of this presentation is to provide the clinical chemist with sufficient information to handle the most common problems he encounters. The treatment given is not, in all cases, rigorous from the statistical standpoint. Standard texts should be referred to for a more thorough consideration in such instances.

Frequency Distributions

If an analysis is performed on a large number of normal people it would be expected that the greatest density of values obtained would occur at, or close to, the arithmetic average or mean of all the values obtained and that the density would decrease steadily as we went away from the average in either direction. We would expect a similar

205

pattern of results if many replicate determinations were made on aliquots of a single blood specimen. In either case, the scatter of results can be depicted graphically by what is called a *frequency curve*. Most of the statistical techniques are based on the assumption that the distribution of values follows the symmetrical, bell-shaped, so-called "normal" or Gaussian curve which is shown in Fig. 1.

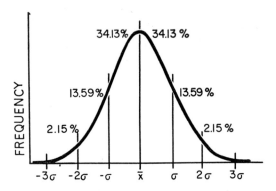

FIG. 1. Distribution of area under the normal curve. The abscissal units are explained in the text.

If the total area under the curve of Fig. 1 is equivalent to all the values plotted, then the area from $-s$ to $+s$ includes about 68% of the values, that from $-2s$ to $+2s$ includes about 95.5% of the values, and that from $-3s$ to $+3s$ includes about 99.7% of the values. The distance, s, in terms of the unit of measurement (e.g., mg. per 100 ml.) is called the *standard deviation* and is a measure of the degree of scatter of the values (the Greek letter sigma, σ, symbolizes the parameter for the whole population, whereas s is the estimate obtained from a sample). The arithmetic average or *mean* $(\bar{x})$ and the standard deviation (s), which completely define any given Gaussian curve, are the two most useful and fundamental indexes in statistical work. Unfortunately, however, valid use of them in many instances is predicated on the "normal," Gaussian frequency distribution which may actually be the exception rather than the rule (1). Indeed, it is perhaps an unfortunate historical fact that the name "normal" was adopted for this distribution. Thus, many distributions encountered are skewed (asymmetrical), platykurtic (too flat), leptokurtic (too peaked), J-shaped, truncated, bimodal, etc. Statisticians have at their disposal certain attacks on the problem of a nonnormal distribution

(e.g., Pearson's family of distributions) but these are complex. On the premise that statistics must be kept simple if they are to be used routinely by chemists it would appear that the "normal" distribution must be usually accepted as a calculated risk. Actually, the risk from a practical standpoint is not as great as has been implied. A first approximation as to whether or not data fit the normal distribution can be made, or, if they do not fit, a simple transformation of the data can frequently be made which will fit. Furthermore, failure of sample data to conform to a "normal" distribution does not necessarily mean that statistical techniques such as tests of significance are invalid. For example, the means from samples of a nonnormal population are approximately normally distributed.

Calculation of the Mean and Standard Deviation

The formula for calculation of the mean, $\bar{x}$, is as follows:

$$\bar{x} = \frac{\Sigma x}{N} \tag{1}$$

where
Σ = sum of the indicated quantities
x = value of an individual observation
N = number of observations

Formulas for calculation of the standard deviation, s, are as follows:

$$s = \sqrt{\frac{\Sigma(\bar{x} - x)^2}{N - 1}} \tag{2}$$

$$= \sqrt{\frac{\Sigma(x^2) - \frac{(\Sigma x)^2}{N}}{N - 1}} \tag{3}$$

where $(\bar{x} - x)$ = the deviation of an observation from the mean, $\bar{x}$. The second form of the equation is more convenient when using tables of squares or a calculator.

Example 1:

Ten replicate determinations ($N = 10$) were made on a single sample of blood.

x	$(\bar{x} - x)$	$(\bar{x} - x)^2$	x^2
10.0	0	0	100.00
10.4	0.4	0.16	108.16
10.1	0.1	0.01	102.01
9.6	0.4	0.16	92.16
9.8	0.2	0.04	96.04
10.1	0.1	0.01	102.01
9.9	0.1	0.01	98.01
10.0	0	0	100.00
10.2	0.2	0.04	104.04
10.0	0	0	100.00
100.1		0.43	1002.43
Σx		$\Sigma(\bar{x} - x)^2$	$\Sigma(x^2)$

$$\bar{x} = \frac{100.1}{10} = 10.0$$

By formula (2):

$$s = \sqrt{\frac{0.43}{10 - 1}} = 0.22$$

By formula (3):

$$s = \sqrt{\frac{1002.43 - \dfrac{(100.1)^2}{10}}{9}} = 0.22$$

Calculations of larger sample data may become awkward without a calculator, unless the work is simplified. This may be done by the following rules, which are stated here without proof (2).

(1) Addition (or subtraction) of a constant, k, to each value of x will increase (or decrease) the mean, $\bar{x}$, to a new value $\bar{x} + k$ (or $\bar{x} - k$) but will not change the variance, s^2 (the symbol s^2, which is the square of the standard deviation, is known as the *variance*).

(2) Multiplication (or division) of each value of x by a constant, k, will multiply (or divide) the mean $\bar{x}$ to a new value $k\bar{x}$ (or to a new value $\bar{x}/k$), *and* will also change the variance s^2 to a new value $k^2 s^2$ (or s^2/k^2). By means of these rules it is possible to reduce or code rather large numbers to others which can be manipulated mentally

or on a slide rule. The mean and variance of the coded data can then be transformed to apply to the original data.

Normal Range

General comments. A normal value of a substance may be defined as the amount of that substance present in a body fluid or excretion of a healthy human being. A "healthy human being" is not so easily defined. In the first place, one is forced to accept current conceptions of the term. Certain variations in values may be regarded as normal today only because of failure to recognize a correlation with an incipient or obscure pathologic process. The establishment of normals is further complicated by many variables, including sex, age, race, climate, season, diet, time of day, day to day variation, menstrual cycle, and whether or not the individual is in the postabsorptive state. The discovery that there is a rhythmic or arrhythmic oscillation of values from day to day in each individual is perhaps not surprising but it certainly further complicates an already complicated picture (3, 4). It becomes obvious that a thorough study of the normal range of any one substance becomes a project of tremendous proportions and it is safe to say that not even one substance has been so treated. Compound these difficulties with the facts that for almost every analysis there are numerous methods available with varying specificities and that for each method there are many modifications given in the literature, and the situation appears almost hopeless. Consider the position of the clinical chemist who must choose from the literature a top normal limit for serum uric acid. Many such examples could be cited and yet the clinical chemist must choose a "normal range" for each analysis he performs.

Regardless of the difficulties involved, there are instances where the clinical chemist must himself establish a normal range. The first question to be answered is what per cent of the total normal population do we wish the normal limits to embrace. To begin with, a dilemma exists in that as the limits are extended to include more of the normals, more abnormals will be included. This is a problem of overlapping distributions as illustrated in Fig. 2. Values to the left of line 1 can, for practical purposes, be considered always abnormal, those to the right of line 2 always normal, but between lines 1 and 2 there is no way of telling from the value per se to which distribution it belongs. Thus, if we include this intermediate area within our

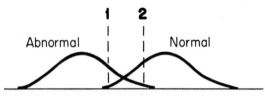

FIG. 2. Overlapping distributions. Values falling between the vertical lines constitute the "zone of suspicion," since they may belong in either of the two distributions.

normal range (taking line 1 as our lower limit of the normal range), a considerable number of abnormals will be judged normal. This is an error of the first kind, that of missing a diagnosis. If, on the other hand, we include the intermediate area with the abnormal distribution (taking line 2 as the lower limit of the normal range), a considerable number of normals will be judged abnormal. This is an error of the second kind, that of classifying a normal person as diseased. Placing the line of demarcation somewhere between positions 1 and 2 would minimize both kinds of error as much as is possible. The proper choice is dependent on the particular test and its purpose but it is generally felt that the error of viewing with suspicion a result that is actually normal is less serious than passing a result that is really pathologic. For this reason it has been proposed (5, 6) that two sets of limits be employed, 80% and 98%. Any value outside the 98% limits would be considered abnormal and those falling between the 80% and 98% limits would be viewed with suspicion. This is theoretically a valid and useful solution to the problem but from a practical standpoint is rather cumbersome and it is doubtful whether it will meet with wide acceptance. A fair compromise is to use 95% limits, i.e., the range in which 19 out of 20 normals will lie. These are by far the most widely used limits and, in fact, are those most widely employed in statistics in general. It must be remembered that, when using a 95% range, approximately 1 normal out of 20 will fall outside *either* limit. Approximately 1 out of 40 will exceed the normal range and 1 out of 40 will fall below the range.

Assuming analysis error to be negligible, the total variation in values obtained on a sample from a normal population is comprised of the sum of "between-individual" variation and "within-individual" variation. It has been suggested (7) that consideration of the "within-individual" variation would be a useful supplement in deciding whether or not a series of values from a given individual indicates normalcy or not. Within-individual variation is usually significantly less than between-individual variation and a given individual normally will fluctuate around his own normal mean. An increase or decrease beyond the normal fluctuation, even though it is still within the "between-individual" normal range, is evidence of an abnormal situation. The two components of normal variation can be evaluated if analyses made for determining normal range are run at two different times on each individual. This concept already has limited application. For example, it is commonly recognized that it is best to have an individual's "baseline" serum cholinesterase value prior to possible exposure to organic phosphate poisoning.

A frequent means of expressing normal limits is to present the total range (the lowest and the highest) obtained for the series of samples studied. This is the least efficient of the methods available since it usually underestimates the 95% limits when obtained from small samples and usually overestimates them from large samples.

If the frequency distribution is of the normal, Gaussian type, the normal range can be characterized by the mean, $\bar{x}$, and the standard deviation, s. A normal range including 95% of the normals can then be taken as $\bar{x} \pm 2s$. On a priori grounds, however, it would be predicted that many distributions of normal values would be skewed but that the distribution of the logarithms of the values would become "normal" (8). Analyses of frequency distributions of blood constituents in normal people have borne out this prediction (6). Biological data in general frequently fall in distributions based on the geometric mean rather than the arithmetic mean. Such a distribution is one which is "normally" distributed after log transformation of the data (a lognormal distribution) and the geometric mean is the antilog of the arithmetic mean of the logs of the observations. A lognormal distribution occurs when an arithmetic change in a cause makes a geometric change in effect. In cases where the normal values extend over less than a 2-fold range, log transformation has little effect on the shape of the distribution. In the case of more extensive distributions which are skewed, however, mathematical calculations of the normal range on the basis of $\bar{x} \pm 2s$ will be in significant error. Examples have actually appeared in the literature where the $\bar{x} - 2s$ yields a negative value for the lower limit (9).

Procedure for calculation of the normal range. The first step is to determine whether or not the data fit the normal or lognormal distributions reasonably well. Visual inspection of data by even a practiced eye usually will not detect a significant deviation from these distributions. Numerous calculational methods exist to answer this question, but as a rule they involve considerable labor (10). A very simple graphic method, which requires only ordinary graph paper, has been described by Moore *et al.* (8). It is called the *normal equivalent deviate (N.E.D.) method.* Observations are arranged in order of increasing magnitude and a log transformation is made if the greatest value is more than twice the least value. Using linear grid graph paper, these values are then plotted on the *x*-axis versus the corresponding N.E.D. values on the *y*-axis (Moore *et al.* (8) give N.E.D. values for all sample sizes up to 100; Table I gives N.E.D. values for sample sizes from 10 to 50 in increments of 5). After all the values are plotted, the best straight line, representing all the points, is drawn using a transparent ruler and visual judgement. The sample is considered normally distributed if the deviations from a straight line are random rather than systematic. If the values appear to describe a curve with the concavity upwards, a straight line (normalcy) might be obtained if the original values without log transformation are plotted versus N.E.D. If neither the original data nor the logarithmic data produce an approximately linear N.E.D. plot the chances are good that the sample is not a uniform one, but is complicated by one or more inhomogeneities. There may be a difference in values obtained in males *vs.* females, or of the techniques of two different analysts, etc. In some cases, the data can be normalized by a "sorting" process, provided the inhomogeneity can be recognized.

The second step is actual calculation of the "95% limits." The procedures discussed here are valid *only* if the data employed in the calculation are approximately "normally" distributed. The $\bar{x}$ and s are calculated by formulas 1 and 2 or 3, respectively. A normal range including 95% of the normals could be calculated as $\bar{x} \pm 2s$ if the values for these two parameters were actually those of the population. In practice, however, one obtains estimates of these from a relatively small sample and the problem is that of estimation of population limits from statistical parameters obtained from a sample taken from the population. This problem is solved for normal distributions by calculation of the so-called "tolerance limits" from the expression $\bar{x} \pm Ks$. (28). Table II gives values of K for calculation of "95% limits" with 90% confidence that the limits so calculated include at least 95% of the population. Tables of K values are also available

(28) that give tolerance limits with 75, 95, and 99% confidence; the choice of 90% confidence is an arbitrary compromise.

In the case of log transformation of data, log x is used for values of x. The $\bar{x}$ and s are then obtained as logs and the "95% limits" of the normal range are then calculated as $\bar{x} \pm Ks$. The antilogs of the upper and lower limits are then taken. The antilog of $\bar{x}$ in this case is the *geometric mean*.

As an alternative to mathematical computation of the $\bar{x}$ and "95% limits" it is possible to take them directly from the N.E.D. plot. The $\bar{x}$ occurs at the intersection of the line drawn at N.E.D. = 0 and the upper and lower limits occur at the intersection of the line drawn and N.E.D. = $+K$ and $-K$, respectively. It is believed, however, that mathematical computation is a more efficient approach if reasonable conformity to normalcy has been established, especially in the case of smaller samples.

Units	Log units	N. E. D.
89	1.950	− 2.054
112	2.050	− 1.555
118	2.073	− 1.282
133	2.125	− 1.080
133	2.125	− 0.915
138	2.140	− 0.772
145	2.162	− 0.643
147	2.168	− 0.524
148	2.171	− 0.413
159	2.202	− 0.306
161	2.207	− 0.202
176	2.246	− 0.100
177	2.248	0.000
180	2.256	0.100
190	2.279	0.202
193	2.286	0.306
198	2.297	0.413
200	2.301	0.524
208	2.318	0.643
218	2.330	0.772
245	2.390	0.915
256	2.409	1.080
267	2.427	1.282
269	2.430	1.555
350	2.544	2.054

Example 2:

It was desired to obtain the normal adult urine levels of amylase
employing a saccharogenic method. Twenty-five random urine specimens were obtained from 7 women and 18 men. There did not appear
to be any difference in the ranges of males and females so the values
were pooled. The values were then listed in order of increasing
magnitude, their logarithms obtained and the N.E.D. values placed
alongside (see table on previous page).

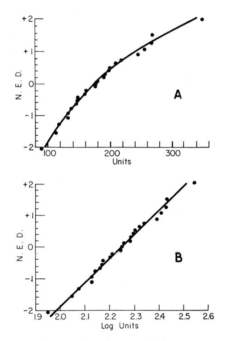

Fig. 3. Plot of normal equivalent deviates (A) versus units and (B) versus
log units.

In Fig. 3A the units are plotted against N.E.D. and it is obvious that
the data cannot be fitted by a straight line. A straight line, however,
adequately fits the data when log units are plotted against N.E.D. as
in Fig. 3B. By taking the antilogs at the points where the line crosses
N.E.D. values of +2.47 and −2.47 (N = 25; K = 2.474), 95% limits
may be estimated. The estimated limits are 83 to 372. The estimated
geometric $\bar{x}$ is 177.

The $\bar{x}$ and "95% limits" are calculated by mathematical computation as follows:

$$N = 25$$
$$\Sigma x = 56.143$$
$$\Sigma(x^2) = 126.525135$$
$$(\Sigma x)^2 = 3152.036449$$
$$\bar{x} = \frac{56.143}{25} = 2.245$$

$$s = \sqrt{\frac{126.525135 \times \dfrac{3152.036449}{25}}{24}} = 0.136$$

95% limits $= 2.245 \pm (2.47)(0.136) = 1.909$ to 2.581. These values for $\bar{x}$ and 95% limits are in terms of logs.
Taking their antilogs:

$$\bar{x} = 176 \text{ units}$$
$$95\% \text{ limits} = 81 \text{ to } 381 \text{ units.}$$

This mean and range is in contrast to those obtained by calculating the mean and range without log transformation: $\bar{x} = 184$ and "95% limits" are 39 to 329.

Statistics of the Analysis

Accuracy. Inaccuracy is defined as the extent to which a measurement differs more or less uniformly from a true value, and is sometimes described as an error of bias. The determination of absolute accuracy is an extremely difficult task, not commonly undertaken by most clinical chemists. Instead, measurements made by the method in question may be compared with those obtained by an acceptable reference method, and the significance of observed differences examined statistically. Many sorts of comparisons can be made, one method against another, one technician against another, one spectrophotometer against another, or even one sample against another. The statistical method for such comparisons is presented under Miscellaneous Statistical Tests, Test of Significance of Difference between Means.

Precision. The precision of an analysis may be established without respect to the accuracy, since precision may be defined as the random error expected of repeated measurements. It is, in other words, the reproducibility of the analysis under the prescribed conditions. Precision is commonly expressed in terms of the standard deviation, s, as previously defined [see Eq. (2) and (3)]. Rather than running many replicate analyses on a single specimen it is usually more feasible to derive an estimate of s from many specimens analyzed in duplicate.

$$s = \sqrt{\frac{\Sigma(d^2)}{N}} \tag{4}$$

where d = difference between duplicates
             N = total number of determinations

As a useful statement of precision, again the "95% limits" are commonly employed. These limits are $\bar{x} \pm ts$, t being obtained from Table III. The "degrees of freedom" (d.f.) are $N - 1$ when using formulas (2) or (3) and $N/2$ when using formula (4). The value t thus can be taken as 2 when d.f. > 30. Having obtained the 95% limits, it might be thought that one can say that, for any single analysis, x, the true result by the method employed has 19 chances out of 20 of being in the range $x \pm ts$. Unfortunately, this is not usually the case. What has been obtained is the "within-run" precision which is usually somewhat better than the reproducibility obtained in a laboratory if variation between days and between analysts is included. An estimate of precision including these sources of variation is a more valid one and, as will be mentioned later, can frequently be taken directly from control charts. This general problem will be discussed in greater detail in the section: The Over-all Error of an Analysis.

Example 3:

(a) *Using formulas (2) or (3).* Taking the data presented in Example 1, the 95% limits are $\bar{x} \pm ts = 10.0 \pm (2.26)(0.22) = 9.5$ to 10.5.

(b) *Using formula (4).* On 12 successive blood determinations each run in duplicate, the following results were obtained:

Results	d	d^2
6.7, 6.9	0.2	0.04
5.8, 5.8	0	0
6.0, 5.6	0.4	0.16

Results	d	d^2
7.1, 7.0	0.1	0.01
6.3, 6.4	0.1	0.01
7.0, 7.6	0.6	0.36
5.9, 6.2	0.3	0.09
6.4, 6.4	0	0
6.4, 6.0	0.4	0.16
6.0, 5.7	0.3	0.09
7.2, 6.7	0.5	0.25
6.3, 6.4	0.1	0.01
$N = 24$		1.18
		$\Sigma(d^2)$

$$s = \sqrt{\frac{1.18}{24}} = 0.22$$

The standard deviation is quite sensitive to sample size (when the sample contains 20 or less observations) and even to the value of the mean. For this reason, stated values of the standard deviation should be accepted only insofar as sample sizes are comparable, and then only in a range close to the mean from which it was determined. These concepts are often combined and expressed as the *coefficient of variation* (C.V.), which is the ratio of the standard deviation to the mean, expressed as a percentage, as

$$C.V. = \frac{100s}{\bar{x}} \qquad (5)$$

The "95% limits" are now $\bar{x} \pm t \, (C.V.)\%$

Example 4:

(a) From Example 3(a)

$$95\% \text{ limits} = 10 \pm (2.26) \left(\frac{(100)(0.22)}{10} \right) \%$$

$$= 10 \pm 4.5\%$$

Thus, the precision (95% limits) of the test at the level of about 10 is $\pm$ 4.5%.

(b) In Example 3(b) the values obtained on the 12 samples ranged from about 5.8 to 7.3 giving an average result of approximately 6.5.

$$95\% \text{ limits} = 6.5 \pm (2.18) \left(\frac{(100)(0.22)}{6.5} \right)$$

$$= 6.5 \pm 7.4\%$$

Thus, the precision (95% limits) of the test in the analysis range of about 5.8 to 7.3 is approximately ± 7.4%.

Effect of replication. The increase of precision, as the average is taken of a greater and greater number of replicates, is given by the following formula:

$$\text{Precision of mean of } N \text{ replicates} = \frac{\text{precision of a single analysis}}{\sqrt{N}} \quad (6)$$

If the 95% limits of a single analysis are ± 10%, then the 95% limits of the averages of 2, 3, and 4 replicates are ± 7, ± 5.8, and ± 5%, respectively. Obviously, this is not a very efficient method of increasing the precision of a result, although frequently it is the only way the desired precision can be achieved. Actually, the improvement in the result may fall short of that indicated by the formula since replicate analyses run at the same time may not be truly random (11). Certainly, replication greater than duplicates, or perhaps rarely triplicates, is totally impracticable in the clinical laboratory. In many instances the chief advantage accruing from running an analysis in duplicate is checking against gross errors such as loss of precipitate upon decanting.

The Over-all Error of an Analysis

The term over-all error embraces the effect of all factors which affect the result of an analysis. Many chemists in an effort to measure the over-all error, have participated in survey programs where a sample is distributed among several laboratories. The results of most such surveys have been disappointing to all parties, and the over-all error has not, in most instances, been examined in terms of its components. If we are speaking about the over-all error of a particular method and eliminate any consideration of specificity (accuracy), there are several component variables which contribute to this total error. There will be a certain variance associated with the replication of any analysis by a given analyst in a given laboratory on a given day. This is equivalent to his within-run precision (P^2). There will also be a variance associated with the data he would obtain on different days (D^2), other conditions being apparently identical. Similarly there will be an element of variance associated with the results of different analysts in the same laboratory (A^2) and an element associated with the results from different laboratories (L^2). Assuming that each of these elements is completely independent of the others, the over-all variance (T^2) is given by the equation

$$T^2 = P^2 + D^2 + A^2 + L^2 \qquad (7)$$

T, P, D, L, and A are in terms of standard deviations or coefficients of variation. In many instances, the only statement about the data available in a given laboratory is the within-run precision, but this can easily be extended to include statements of the effects of factors such as A and D in the equation above. It is then possible to "certify" the precision of an analysis with due regard for factors represented by D and A. It is obviously a more difficult matter to account for the factor L.

The same line of reasoning can be extended to other situations. A method generally includes such operations as deproteinization (D), taking an aliquot (A), adding a reagent (R), diluting to volume (V), and photometry (P). There is distinct variance associated with each and every operation which contributes to the total variance (T^2) as

$$T^2 = D^2 + A^2 + R^2 + V^2 + P^2 \qquad (8)$$

If the experiment is properly designed, it is possible to determine the magnitude of each of these sources of variance and whether or not any differences in magnitude are significant. The statistical technique employed is called *analysis of variance* (12, 13, 14, 15). Complete details will be found in the references cited, but a brief example will serve to demonstrate the power and the usefulness of analysis of variance.

Example 5:

To examine the quality of serum sodium analyses, four different samples, A, B, C, D, were selected. These were examined, for reasons of symmetry, on four different days by four different analysts, *according to a preconceived plan*. Each analyst reported the observed value. To simplify the calculations, each value was "coded" by subtraction of the arbitrary constant, 120 mEq./1. This does not affect the subsequent statistical analysis. The data were then arranged as in the following table.

		Day		
Analyst	1	2	3	4
1	A–13	B–15	C–10	D–11
2	B–10	D–12	A–9	C–7
3	C–10	A–12	D–12	B–11
4	D–13	C–12	B–10	A–12

Note that according to the chosen plan each sample, represented by a letter, appears once in each row and once in each column. In other words, each sample was analyzed once by each analyst and once each day. (The above array of data is known as a 4×4 Latin square, and was chosen at random from the 576 possible 4×4 Latin squares (14). From the preliminary array we need to know the sum of each row, of each column, and the sum of all values for each sample considered without regard to position. We shall also need the sum of squares of all values, $\Sigma(x^2)$, and the sum of all the values (T). Since the rows in the original array corresponded to analysts and the columns to days, we shall describe the calculations accordingly. We must compute the following sums of squares of deviations from the $\bar{x}$:

$$\text{Total sum of squares} = \Sigma(x^2) - \frac{(\Sigma x)^2}{N}$$

$$= 2055 - \frac{(179)^2}{16}$$

$$= 52.44$$

$$\text{Analyst sum of squares} = \frac{\Sigma \text{ (subtotals by analyst)}^2}{\text{no. observations per analyst}} - \frac{(\Sigma x)^2}{N}$$

$$= \frac{(49^2) + (38^2) + (45)^2 + (47)^2}{4} - 2002.56$$

$$= \frac{8079}{4} - 2002.56 = 17.19$$

$$\text{Days sum of squares} \quad = \frac{\Sigma \,(\text{subtotals by days})^2}{\text{no. observations per day}} - \frac{(\Sigma x)^2}{N}$$

$$= \frac{(46)^2 + (51)^2 + (41)^2 + (41)^2}{4} - 2002.56$$

$$= \frac{8079}{4} - 2002.56 = 17.19$$

$$\text{Samples sum of squares} = \frac{\Sigma \,(\text{subtotals by sample})^2}{\text{no. observations per sample}} - \frac{(\Sigma x)^2}{N}$$

$$= \frac{(46)^2 + (46)^2 + (39)^2 + (48)^2}{4} - 2002.56$$

$$= \frac{8057}{4} - 2002.56 = 11.69$$

We can tabulate the further calculations in the manner shown below.

	Sum of squares	d.f.	Mean squares or variances	F ratio	Significance
Total	52.44	15			
Analysts	17.19	3	5.73	5.41	+
Days	17.19	3	5.73	5.41	+
Samples	11.69	3	3.89	3.67	—
Residual	6.37	6	1.06		

Since the total sum of squares is related to the total variance, and since the row, column, and sample sums of squares are related to the variance elements due to the arrangement of the data, we must still seek a measure of the variance which has been "washed clean" of any arrangement element. In terms of the example, we seek a measure of the variance which is independent of the effects of the analysts, of the day on which the analysis was made, or of the samples being analyzed. This, we find in the *residual* sum of squares, which is obtained by subtracting from the total sum of squares all the other calculated sums of squares. When divided by the proper degrees of freedom, each sum of squares is reduced to a mean square or variance as shown in the table. The total degrees of freedom is $N - 1$. The degrees of freedom for analysts, days, and samples is, in each case, one less than the number of analysts, days, and samples involved. The "residual" degrees of freedom is the total less the sum of those for analysts, days, and samples. The final step consists of dividing each pertinent mean square by the residual mean square to obtain F ratios which express the significance of the elements of arrangement. If the calculated F values exceed the tabulated values at any predetermined level we may conclude that the effect is of significance. From Table IV, we learn that the critical value of F for the 95% limits, with 3 degrees of freedom in the numerator and 6 degrees of freedom in the denominator, is 4.76. In the present example, we conclude, at the 95% level, that there is no significant difference in the samples, but that there is a significant difference in the performance of the four analysts who checked them. Furthermore, there is a significant difference in the quality of the work performed on different days.

The simple case outline above involved two variables of classification (days and analysts) with only a single observation in each cell of the Latin square. That element of the variance which was independent of the classification was termed the residual variance. With only single observations in each cell, no further analysis is possible. If, however, the experiment is modified to include more than one observation in each cell it becomes possible to also examine the variance of the items in the individual cells. This is sometimes termed the "within-group" variance. The remaining element of variance is termed the element of *interaction*. In the specific sense of its present use, the term interaction describes effects produced by the mutual action of two variables which would not be produced by either considered separately.

If, for example, the experiment described above was repeated in each of three separate periods, the analysis could be made separately for each. Alternatively, the replicates could be pooled, cell by cell, and the entire program regarded as a single replicated experiment. The latter case could be examined for interaction, which in the example discussed above could be ascribed to some unrecognized effect of the day on the analyst, or vice versa.

In summary, analysis of variance, given a proper experimental design, permits the simultaneous study of more than two factors which affect the over-all error without the necessity of holding any factor constant. It represents an extension of the *t*-test which can extract the maximum amount of information from a given quantity of data, provided the data are gathered according to plan. It should be emphasized that the present discussion is intended as an outline of basic principles, not as a complete discussion. The cited references should be consulted for full details of computation and interpretation.

Claims are frequently made in the literature (16, 17) that a given method has a reproducibility of ± 1% or ± 2%. These claims may or may not apply under ideal conditions, but many such statements are unrealistic in terms of day-to-day working conditions and performance (18, 19, 20, 21). The analysis of variance may seem to throw a gloomier light on the picture of error, but it is a far more honest light. It has the further advantage of pointing out the contributory factors of major importance, so that some effort may be made to improve the picture. Obviously, the nature and extent of the errors should be reinvestigated whenever the analysts or the methods are changed.

Practical aspects of laboratory error. In designing the routine procedures of a laboratory one must stay within the bounds of practicality and still obtain the accuracy and precision desired. Valuable time in the laboratory is wasted by complicating certain procedures to give greater accuracy and precision than is actually required. That this requirement can vary considerably can be demonstrated by the following consideration. If one accepts 4.1 to 5.6 mEq./1. as the normal range of serum potassium, then ± 16% from the middle spans the entire range. If one accepts 131 to 150 mEq./1. as the normal range of serum sodium, then ± 7% from the middle spans the range. It would seem reasonable that the precision requirements should be more stringent for sodium than for potassium. It is the chemist's responsibility to see that these requirements are met. As already noted, interlaboratory accuracy surveys have revealed that most determinations are not meeting any reasonable standards of tolerance. Placing the blame on the analysts is not necessarily the answer, for some methods as they exist may be incapable of rendering results which will fall within the maximal permissible tolerances. The authors regard the control of accuracy and precision as the most pressing problem existing in the field of clinical chemical technology today.

Control of Laboratory Error

To the responsible chemist, the control of error is no less important than the measurement of error. He must seek assurance that unforeseen errors do not arise and that the errors of which he is aware are held at a reasonably constant level.

Standards. It goes without saying that a first requirement is the standardization of every method used by the laboratory; no chemist worthy of the title should rely on "precalibrations" of any sort. Initial standardization is not enough, however, since changes can and often do occur some time after the original standardization. For this reason, standards should ideally be included with every set of analyses, or at least once a day. Less frequent standardization entails a certain risk, acceptance of which is a decision that can soundly be made only after considerable experience *with* daily standardization. There seems to be little doubt that failure to frequently and properly standardize methods is the largest single factor contributing to the poor performance seen in many surveys.

There are three types of standards generally available. The first is a *pure standard.* This consists of a solution of the substance under study in a known state of purity, perhaps with a preservative added. Any other type of standard must ultimately be referred to a pure standard. With few exceptions, pure standards are available, for the most part from crystalline substances. In some instances, the very purity which recommends them is a drawback, since they lack possible inhibitors or contaminants which may occur in the routine samples to be analyzed. Although many laboratories do employ pure standards, in some instances they are employed improperly. As far as possible, standards of any sort should always be run through the entire analytical procedure. Omission of the deproteinization steps, for example, may permit a contaminated or decomposed reagent to escape detection.

A second type of standard is the *internal standard,* in which a known amount of a pure standard is added to an aliquot of the routine sample. This technique could be more widely used in routine analysis than at present. It is most suited to the measurement of recovery, but has wider applications. The quantity added must, of course, be greater than the limiting sensitivity of the method, yet the final concentration should fall well within the linear portion of the calibration curve. Internal standardization also requires that the "spiked" sample be subjected to the entire analytical procedure.

The third type of standard takes the form of a *pooled sample.* It is simple to collect and pool submitted samples of plasma, serum, or even urine. The pool is split into small aliquots some of which are then carefully analyzed. The remainder may then be stored, preferably at $-10°C$. or lower. Many constituents of serum are stable for at least 6 months under these conditions (22). The samples may be dried in the frozen state, and many commercial standards are now available in lyophilized form. Frozen or lyophilized samples have much to recommend them, but certain precautions are vital, particularly with the lyophilized material. Some analysts forget that the stated value for a concentration in a commercial sample applies only to the exact method used in their standardization, which is not necessarily identical with the method he may be using. It is also true that the reconstitution of a lyophilized sample may introduce significant error. This involves not only the volume of the reconstituting fluid added, but also to the manner in which the dried material is dispersed. Shaking violently will do considerable damage to some enzymes; lyophilization does not confer indefinite stability, and such standards should always be provided with a definite date of expiration. This is especially true of standards made of dried Red Cross plasma. Standards of this material are particularly unstable to violent shaking, so that even the PBI may be affected (23).

Quality control charts. Given a standard of the chemist's choice, there is a simple and effective way to use the accumulated data. They can be used as the basis of a *quality control chart*, which visually depicts how far the measured values may vary before suspicion of the data is warranted. Control charts can also be used to warn of trends, which, if left unchecked, may become significant errors. There are several ways in which control charts may be constructed, of which two will be discussed here.

Single standard charts. Single standards are run and plotted on

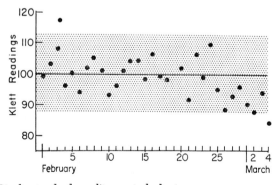

Fig. 4. Single standard quality control chart.

226 HENRY AND DRYER

graph paper as shown in Fig. 4. The observed values of the standards, either in terms of final concentration or in terms of some measure such as an absorbance, are plotted on the ordinal scale. At least 20 such standard values should be obtained as soon as possible. The mean, $\bar{x}$, of these observed values is then calculated, and a line drawn through the value of $\bar{x}$ should be extended to the right where future values will be entered in the chart. Tolerance limits (usually 95% limits) are then established for the plotted points, with lines at the limiting values drawn on the chart equidistant from the line through the mean. These limit the area on the chart in which successive standard analyses should fall about 95% of the time, assuming that no factors are operating which did not exist during the period in which the original 20 observations were collected. For reasons discussed below, the 95% limits are generally taken as $\pm$ 3s in control charts, rather than the usual $\pm$ 2s. If all goes well and the accumulation of points remains within the established belt, $\bar{x}$ and s may be recalculated using all the collected data.

The value of s can be determined by formulas 2 or 3, or by treating successive determinations of standards as pairs, i.e., x_1 and x_2 as a pair, x_2 and x_3 as a pair, etc. The differences between pairs, R, and then their average, $\bar{R}$, are calculated. Limits of 2s and 3s are then calculated by 1.77 $\bar{R}$ and 2.65 $\bar{R}$, respectively. This method of calculating s usually yields a close approximation to s calculated by the standard formula.

There are always two dangers when using control charts, (1) looking for error that does not really exist, and (2) not seeing error that does exist. As tolerance limits are increased, errors of the first kind are decreased and those of the second kind are increased. A safe compromise seems to be the use of $\pm$ 3s as the 95% limits until experience indicates that this exceeds the desired precision, in which case $\pm$ 2s limits may be imposed on the chart. In this instance, out-of-limits points will occur, from time to time, in the absence of demonstrable sources of error.

What course of action should be taken when a value falls outside of the belt representing the tolerance limits? It should be kept in mind that about 1 in 20 values *should* fall out of limits, but not very far outside. When this occurs, the entire method should be carefully examined for an assignable cause, e.g., a new reagent, a new standard, a change in technique of some sort, or a change in glassware and apparatus. If none of these is discovered and if the point was not too far removed from the limit, the best approach is to wait to see where the next point falls. If the excessive deviation persists there is due cause for alarm. In many cases trouble can be anticipated before the points actually fall outside the belt. In Fig. 4, it is noted that after Feb. 26 there is a gradual downward trend of the successive points. Apparently some difficulty began about that time but did not become significant until about March 4. This is the type of drift that generally indicates a gradual deterioration of a reagent or of a standard.

The control chart limits constitute a very good estimate of the between-run precision of an analysis if the values plotted are the results obtained on a pooled serum or urine.

Duplicate standard charts. By running standards in duplicate it is possible to tell if the variation between runs is significantly greater than the variation in any one run. With many chemical procedures this may be the case, indicating insufficient control of such variables as temperature. The mean, $\bar{x}$, of each pair and the differences between

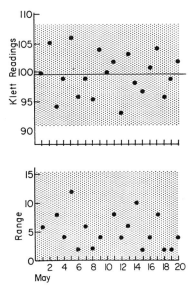

FIG. 5. Duplicate standard quality control chart. This chart also shows the range of the duplicate standards.

each pair (the range, R) are plotted as shown in Fig. 5; $2s$ and $3s$ limits for the $\bar{x}$ are $\pm$ 1.26($\bar{R}$) and $\pm$ 1.88($\bar{R}$), respectively, where the value of $\bar{R}$ is obtained by averaging the range values, R. This is done after about 20 pairs are accumulated. The upper $2s$ and $3s$ confidence limits for the range, R, are $2.18(\bar{R})$ and $3.27(\bar{R})$, respectively. The lower limit is 0 in either case. When the R values are in control (within the confidence limits) but the averages of duplicates are out of control (outside the confidence limits), the interpretation is that the variation between runs is significantly greater than that within runs.

Henry and Segalove (18) have recommended that the control chart mean, $\bar{x}$, be used for the value of the standard in calculating unknowns, not the value for the standard obtained in the same run with the unknowns. Benenson et al. (24) have taken issue with this suggestion, maintaining that use of the standard value obtained in the same run with the unknowns corrects for small variations due to factors such as time and temperature. Logically, the former viewpoint is correct if there is no correlation between the errors of standards and unknowns whereas the latter viewpoint is correct if there is complete correlation. Actually, it is probable that in most instances there is partial correlation. At the present time, it is not possible to say which standard value is better to use simply because a statistical analysis of this rather complex problem has not as yet been made. The following recommendations are, therefore, tentatively made: (1) Employ the control chart mean if (a) standards are run in single, or, (b) if standards are run in duplicate and there is no evidence of greater variation between runs than within runs (standards and ranges in control). (2) Employ the average of the pair of standard values obtained in the same run with the unknowns if the ranges are in control but the standards out of control.

Reagent blanks. In most analyses, a reagent blank is run on the materials used in the determination. The most common procedure in photometric analyses is to set the photometer at 100% transmission on the reagent blank, thus achieving an automatic correction for the blank in the readings of the unknown and standard. Just as for unknowns and standards, however, there is error connected with the measurement of the blank. The error contributed by variation of the blank can be reduced by reading the standards, unknowns, and blanks against distilled water (or the solvent used), which is a fixed reference of absorbance, and then correcting the readings for the blank, using the average of many blanks (say 15 or 20) run over a period of time as the "best estimate" of what the blank should be (18). This approach is feasible only when the absorbance of the reagent blank against water or solvent is fairly low. The values as obtained can be placed on the control chart for the standards. No confidence limits are readily available but a visual record is provided. An increase in the reagent blank signifies deterioration or contamination of a reagent. Reagent contamination is not self-evident when standards are read against reagent blanks.

Significance of the Data

No discussion of error or of statistics would be complete without mention of the significance of numbers. When a concentration is reported as 12.5 mg. per 100 ml. each of the digits is assumed to be significant, with a maximal uncertainty of 1 in the last place. The actual concentration, as far as the report itself is concerned, may be said to lie between 12.45 and 12.55 mg. per 100 ml., so that the implied uncertainty of the report is about 0.4%. If the method itself has an inherent error of ± 10%, the report described above contains insignificant figures, and it should more properly be reported as 12 mg. per 100 ml. If the method employed in the example cited above possessed an inherent error of ± 3%, then one would appear to be justified in retaining the third significant figure since the actual precision of the result is significantly better than that implied by only two significant figures. With the exception of this compromise, all reports should be rounded off to include only figures of significance. Certain general rules have been accepted for this purpose, which may be briefly stated as follows (25). Discard all numbers to the right of the last significant (or nth) place, but if the discarded numbers are more than half a unit in the nth place, increase only *odd* digits.

The above remarks apply especially to reported data, as issued to the referring physician. In handling data to be used in statistical analysis, it is common practice to retain one more place in each item than is desired in the final computations. Rounding off is performed as a last step.

Miscellaneous Statistical Tests

F test for comparison of precisions. This test can be used to detect whether a significant difference in reproducibility exists between two analysts using the same method, two pieces of equipment, two different methods, etc. (It was used earlier in the example of analysis of variance.) The variance, the square of the standard deviation, is calculated for each set of data. Then,

$$F = \frac{larger\ of\ the\ variances}{smaller\ of\ the\ variances} \tag{9}$$

The larger of the two variances is placed in the numerator except in the instance where a priori knowledge dictates that one variance cannot be less than the other. In this case, where $s_1^2 \geq s_2^2$, s_1^2 is placed in the numerator whether its estimate is larger, equal or smaller than s_2^2. To obtain the limiting value of F for any particular probability of significance, enter a table of F (Table IV gives values of F for 95% probability) taking the "degrees of freedom" for the numerator and denominator as one less than the number in the respective sample. If the F value obtained for the two samples exceeds that found in the table of F, the precision of the data corresponding to the numerator is significantly poorer than that corresponding to the denominator.

If the standard deviations are calculated from duplicate analyses on different samples, the table of F is entered with "degrees of freedom" equal to the number of pairs of duplicates in each series.

Example 6:

Twenty blood glucose analyses were performed in duplicate, 10 by analyst A and 10 by analyst B.

A	d	d^2	B	d	d^2
100, 104	4	16	81, 76	5	25
100, 94	6	36	102, 104	2	4
120, 121	1	1	71, 70	1	1
75, 73	2	4	110, 115	5	25
81, 83	2	4	111, 108	3	9
74, 77	3	9	82, 83	1	1
99, 99	—	—	72, 76	4	16
110, 116	6	36	91, 84	7	49
68, 72	4	16	98, 95	3	9
130, 126	4	16	120, 127	7	49
$\Sigma(d^2)$		138			188

$$s_A{}^2 = \frac{\Sigma(d^2)}{N} = \frac{138}{20} = 6.9$$

$$s_B{}^2 = \frac{188}{20} = 9.4$$

$$F = \frac{s_B{}^2}{s_A{}^2} = \frac{9.4}{6.9} = 1.36$$

Entering the table of F (Table IV) with 9 degrees of freedom for the numerator and denominator, the F value is 3.18. Since the F value obtained from the data is only 1.36 it can be concluded that the reproducibility obtained by analyst B is not significantly poorer than that obtained by analyst A.

Test of significance of difference between two means (t test). The clinical chemist frequently is faced with the following problems: (1) Are the values in this set of data significantly higher than the values in the other set? For example, is the normal range for males higher than the normal range for females? (2) Does method A yield, on the average, the same, higher, or lower results than method B? Such questions can be resolved by application of the t test. The comparison is made by calculating, from formulas given below, a statistic known as "t," which is the ratio of the difference between two means to the standard deviation of this difference. Theoretical values of t have been tabulated for several limits of significance, so that the test consists of comparing the value calculated from the data with the theoretical value from the tables. If the calculated value exceeds the tabular value, the difference in the means is significant at the limit chosen. The usual limits are 95% or, less frequently, 99%. These are the points which will include 95 out of 100 and 99 out of 100 cases, respectively.

For paired observations, t is given by

$$t = \frac{\bar{d}}{s_d} \sqrt{n} \quad (\text{"degrees of freedom"} = n - 1) \quad (10)$$

where $n =$ number of pairs of observations
$d =$ difference between the observations of a given pair
$\bar{d} =$ the mean of the differences
$s_d =$ the standard deviation of the differences

$$= \sqrt{\frac{\Sigma(d^2) - \frac{(\Sigma d)^2}{n}}{n - 1}} \tag{11}$$

The critical value of t (95% limits) is obtained from Table III by entering with the proper "degrees of freedom" which is $n - 1$. If the t value obtained from the data exceeds the critical value, the mean difference, $\bar{d}$, is significantly greater than 0.

Example 7:

Sample	Method A	Method B	d	d^2
1	4.9	4.7	0.2	0.04
2	6.0	5.5	0.5	0.25
3	3.1	3.2	−0.1	0.01
4	5.0	4.6	0.4	0.16
5	4.1	3.6	0.5	0.25
6	5.0	4.9	0.1	0.01
7	4.2	4.0	0.2	0.04
8	4.8	4.2	0.6	0.36
9	4.1	3.8	0.3	0.09
10	4.4	4.0	0.4	0.16
Σ	45.6	42.5	3.1	1.37

$$\bar{d} = \frac{3.1}{10} = 0.31$$

$$(\Sigma d)^2 = (3.1)^2 = 9.61$$

$$s_d = \sqrt{\frac{1.37 - \frac{9.61}{10}}{9}} = 0.21$$

$$t = \frac{0.31}{0.21} \sqrt{10} = 4.9$$

Since the value for t for 9 degrees of freedom equals 2.26, one can conclude that there is a greater than 95% probability that the results yielded by method B are significantly lower than the results by method A.

For unpaired observations, the following formula pertains:

$$t = \frac{\bar{x}_1 - \bar{x}_2}{\sqrt{s_{\bar{x}_1}^2 - s_{\bar{x}_2}^2}} \tag{12}$$

where $\bar{x}_1$ and $\bar{x}_2$ = means of sets of data 1 and 2, respectively, and

$$s_{\bar{x}_1}^2 = \frac{s_1^2}{N_1}$$

$$s_{\bar{x}_2}^2 = \frac{s_2^2}{N_2}$$

N_1 and N_2 = number of observations in sets 1 and 2, respectively

degrees of freedom = $N_1 + N_2 - 2$

Example 8:

Values were obtained on 12 normal males and 16 normal females with the following results

$\bar{x}_M = 63$

$s_M = 8$

$\bar{x}_F = 51$

$s_F = 6$

Is there a significant difference between the males and females?

$$t = \frac{63 - 51}{\sqrt{\frac{8^2}{12} + \frac{6^2}{16}}} = 4.37$$

The critical level (95% limits) for t with 26 d.f. is 2.05, hence, there is better than 95% probability that males are higher than females. Reference to a more complete table of t reveals that the 99.9% limits for 26 d.f. is 3.707, hence there is even better than 99.9% probability that males are higher than females.

Conclusion

Chemists must become as critical of their data as they are of their reagents. The techniques described above, admittedly elementary by mathematical standards, can be of considerable assistance in control and elimination of error in routine laboratory performance.

REFERENCES

1. Clancey, V. J., Statistical methods in chemical analyses. *Nature* **159**, 339 (1947).
2. Dixon, W. J., and Massey, F. J., "Introduction to Statistical Analysis." McGraw-Hill, New York (1951).

234 HENRY AND DRYER

3. Margolese, M. S., and Golub, O. J., Daily fluctuation of the serum protein-bound iodine level. *J. Clin. Endocrinol. Metab.* 17, 849 (1957).
4. Elkington, J. R., and Danowski, T. S., "The Body Fluids." Williams and Wilkins, Baltimore, Maryland (1955).
5. Mainland, D., Statistics in medical research. *Methods Med. Res.* 6, Sect. III (1954).
6. Wootton, I. D. P., and King, E. J., Normal values for blood constituents. *Lancet i*, 470 (1953).
7. Schneider, A. J., Some thoughts on normal, or standard, values in clinical medicine. *Pediatrics* 26, 973 (1960).
8. Moore, F. J., Cramer, F. B., and Knowles, R. G., "Statistics for Medical Students." Blakiston, Philadelphia, Pennsylvania (1951).
9. Henry, R. J., Improper statistics characterizing the normal range, *Am. J. Clin. Pathol.* 34, 326 (1960).
10. Dryer, R. L., "Elementary Statistics, Practical Data Analysis." State University of Iowa, Iowa City, Iowa (1956).
11. Moran, R. F., Determination of the precision of analytical control methods, *Anal. Chem.* 15, 361 (1943).
12. Youden, W. J., "Statistical Methods for Chemists." Wiley, New York (1951).
13. Cochran, W. G., and Cox, G. M., "Experimental Designs." Wiley, New York (1950).
14. Federer, W. T., "Experimental Design." Macmillan, New York (1955).
15. Snedecor, G. W., "Calculation and Interpretation of Analysis of Variance and Co-Variance." Iowa State College Press, Ames, Iowa (1934).
16. Bodansky, M., and Bodansky, O., "Biochemistry of Disease." Macmillan, New York (1952).
17. Fischl, J., Determination of protein-bound iodine in micro amounts of serum or plasma. *Clin. Chim. Acta* 1, 462 (1956).
18. Henry, R. J., and Segalove, M., Running of standards in clinical chemistry and the use of the control chart. *Am. J. Clin. Pathol.* 5, 305 (1952).
19. Wootton, I. P. D., Standardization in clinical chemistry. *Clin. Chem.* 3, 401 (1957).
20. Levey, S., and Jennings, E. R., Use of control charts in the clinical laboratory. *Am. J. Clin. Pathol.* 20, 1059 (1950).
21. Henry, R. J., Berkman, S., Golub, O. J., and Segalove, M., Analysis of error in clinical chemical determinations and pipetting procedures. *Am. J. Clin. Pathol.* 23, 285 (1953).
22. Walford, R. L., Sowa, M., and Daley, D., Stability of protein, enzyme, and non-protein constituents of stored frozen plasma. *Am. J. Clin. Pathol.* 26, 376 (1956).
23. Dryer, R. L., unpublished observations.
24. Benenson, A. S., Thompson, L. K., and Klugerman, M. R., Application of laboratory controls in clinical chemistry. *Am. J. Clin. Pathol.* 25, 575 (1955).
25. Nielsen, K. L., "Methods in Numerical Analysis." Macmillan, New York (1956).
26. Fisher, R. A., "Statistical Tables for Research Workers." Oliver and Boyd, Edinburgh (1950).

27. Burington, R. S., and May, D. C., "Handbook of Probability and Statistics With Tables." Handbook Publishers, Sandusky, Ohio (1953).
28. Bowker, A. H., In "Techniques of Statistical Analysis," (C. Eisenhart et al., eds.) pp. 97–110. McGraw-Hill, New York (1947).

APPENDIX I
ABBREVIATED TABLES

These tables are too abbreviated for general use, and are included to explain the discussion in the body of the paper. Complete and exact tables are available in several sources (26, 27).

TABLE I
NORMAL EQUIVALENT DEVIATES

First rank the data in order of magnitude, then associate with each item the proper N. E. D. Read the table downward with negative sign, then upwards with positive sign. These tables are modified from Moore et al. (8); more extensive tables may be found in ref. (8).

10	15	20	25	30	35	40	45	50
1.645	1.834	1.960	2.054	2.128	2.189	2.241	2.287	2.326
1.036	1.282	1.440	1.555	1.645	1.719	1.781	1.834	1.881
0.675	0.967	1.150	1.282	1.383	1.465	1.534	1.593	1.645
0.385	0.728	0.935	1.080	1.192	1.282	1.356	1.420	1.476
0.126	0.524	0.755	0.915	1.036	1.133	1.213	1.282	1.341
	0.341	0.598	0.772	0.903	1.006	1.092	1.164	1.227
	0.168	0.454	0.643	0.784	0.894	0.984	1.061	1.126
	0.000	0.319	0.524	0.675	0.792	0.887	0.967	1.036
		0.189	0.413	0.573	0.697	0.798	0.882	0.954
		0.063	0.306	0.477	0.609	0.714	0.803	0.878
			0.202	0.385	0.524	0.636	0.728	0.806
			0.100	0.297	0.444	0.561	0.657	0.739
			0.000	0.210	0.366	0.489	0.590	0.675
				0.126	0.291	0.419	0.524	0.613
				0.042	0.217	0.352	0.462	0.553
					0.144	0.286	0.400	0.496
					0.072	0.221	0.341	0.440
					0.000	0.157	0.282	0.385
						0.094	0.225	0.332
						0.031	0.168	0.279
							0.112	0.228
							0.056	0.176
							0.000	0.126
								0.075
								0.025

TABLE II
FACTORS K FOR CALCULATION OF NORMAL RANGE,
95% LIMITS, 90% CONFIDENCE

Reference (28) gives complete tables of K for N 2–1000, confidence coefficients of 75, 90, 95, and 99%, and for 75, 90, 95, 99, and 99.9% limits. (This table is reproduced with the permission of the author and the McGraw-Hill Book Co.)

N	K factors
5	4.152
6	3.723
7	3.452
8	3.264
9	3.125
10	3.018
11	2.933
12	2.863
13	2.805
14	2.756
15	2.713
16	2.676
17	2.643
18	2.614
19	2.588
20	2.564
25	2.474
30	2.413
35	2.368
40	2.334
45	2.306
50	2.284
60	2.248
70	2.222
80	2.202
90	2.185
100	2.172

TABLE III
CRITICAL VALUES OF t, 95% LIMIT

d.f.	t	d.f.	t	d.f.	t
1	12.71	11	2.20	21	2.08
2	4.30	12	2.18	22	2.07
3	3.18	13	2.16	23	2.07
4	2.78	14	2.14	24	2.06
5	2.57	15	2.13	25	2.06
6	2.45	16	2.12	26	2.06
7	2.37	17	2.11	27	2.05
8	2.30	18	2.10	28	2.05
9	2.26	19	2.09	29	2.04
10	2.23	20	2.09	30	2.04

TABLE IV
CRITICAL VALUES OF F, 95% LIMIT

d.f. of denom-inator	d.f. of numerator												
	1	2	3	4	5	6	7	8	9	10	15	20	30
1	161	200	216	225	230	234	237	239	241	242	246	248	250
2	18.5	19.0	19.2	19.2	19.3	19.3	19.4	19.4	19.4	19.4	19.4	19.4	19.5
3	10.1	9.55	9.28	9.12	9.01	8.94	8.89	8.85	8.81	8.79	8.70	8.66	8.62
4	7.71	6.94	6.59	6.39	6.26	6.16	6.09	6.04	6.00	5.96	5.86	5.80	5.75
5	6.61	5.79	5.41	5.19	5.05	4.95	4.88	4.82	4.77	4.74	4.62	4.56	4.50
6	5.99	5.14	4.76	4.53	4.39	4.28	4.21	4.15	4.10	4.06	3.94	3.87	3.81
7	5.59	4.74	4.35	4.12	3.97	3.87	3.79	3.73	3.68	3.64	3.51	3.44	3.38
8	5.32	4.46	4.07	3.84	3.69	3.58	3.50	3.44	3.39	3.35	3.22	3.15	3.08
9	5.12	4.26	3.86	3.63	3.48	3.37	3.29	3.23	3.18	3.14	3.01	2.94	2.86
10	4.96	4.10	3.71	3.48	3.33	3.22	3.14	3.07	3.02	2.98	2.85	2.77	2.70
15	4.54	3.68	3.29	3.06	2.90	2.79	2.71	2.64	2.59	2.54	2.40	2.33	2.25
20	4.35	3.49	3.10	2.87	2.71	2.60	2.51	2.45	2.39	2.35	2.20	2.12	2.04
30	4.17	3.32	2.92	2.69	2.53	2.42	2.33	2.27	2.21	2.16	2.01	1.93	1.84

URIC ACID*

Submitted by: WENDELL T. CARAWAY, Flint Medical Laboratory, and Laboratories of McLaren General Hospital and St. Joseph Hospital, Flint, Michigan

Checked by: PAULINE M. HALD, Grace-New Haven Community Hospital, New Haven, Connecticut

Introduction

The colorimetric determination of uric acid in blood is usually based on the reduction of phosphotungstate by uric acid in alkaline solution. The reaction is not specific for uric acid. Under the conditions employed, glutathione and ergothioneine from the red cells account for most of the interference; consequently, determinations are done preferably on plasma or serum. In addition, uric acid values on whole blood will vary with the hematocrit inasmuch as the concentration of uric acid in the red cells is only approximately 55% of that in the plasma (1).

In the original method of Folin and Denis (2) sodium carbonate was used to adjust the alkalinity of the final solution. Precipitation of slightly soluble sodium or potassium phosphotungstates often resulted in turbid solutions. Substitution of lithium oxalate for potassium oxalate as anticoagulant eliminated much of the turbidity. The introduction of cyanide ion into the reaction medium (3) greatly increased the sensitivity of the method. Disadvantages include the poisonous nature of cyanide; the instability of cyanide solutions and the development of considerable color in blank controls. Substitution of spectrophotometric for visual measurements of color in clinical laboratories has favored a return to less sensitive procedures in an attempt to eliminate the need for cyanide (4, 5, 6, 7, 8, 9).

Comparative results obtained by Lous and Sylvest (10) on serum by three different methods of analysis and by Kanabrocki et al., (11) by five different methods have been reported. In the following procedure, solutions of sodium carbonate and phosphotungstic acid are added directly to a protein-free filtrate of serum to develop a blue color proportional to the concentration of uric acid (6). The final concen-

* Based on the method of Folin and Denis (2) as modified by Caraway (6).

239

tration of phosphotungstic acid is sufficiently low to prevent development of turbidity. The concentration of sodium carbonate provides a
pH of approximately 10.2 in the final reaction mixture. This results in
maximum buffering action and optimum color development. A relatively constant plateau of color density is obtained between 30 and 50
minutes after mixing.

Reagents

NOTE: All reagents and standard solutions should be prepared with chlorine-
free distilled water (12).

1. Phosphotungstic acid reagent, Folin and Denis (2). Use a thoroughly cleaned 1000-ml. round-bottomed flask fitted with a ground-
glass joint and equipped with a reflux condenser. In the flask dissolve
50 g. of molybdate-free sodium tungstate, $Na_2WO_4 \cdot 2H_2O$, in 400 ml. of
water. Add 40 ml. of 85% o-phosphoric acid, attach the condenser, and
boil gently for 2 hours. Cool to room temperature, transfer with rinsing
to a 500-ml. volumetric flask and dilute to the mark with water. The
reagent is stored in a brown bottle. It is stable indefinitely and should
be practically colorless. Inasmuch as the acidity of the reagent is
important in the reaction, other phosphotungstic acid reagents may not
be substituted indiscriminantly.

2. Dilute phosphotungstic acid. Dilute 10.0 ml. of Folin-Denis
reagent to 100 ml. with water. This is stored in a brown bottle and
is stable for at least 1 year.

3. Sodium carbonate solution, 10%. Dissolve 100 g. of anhydrous
Na_2CO_3 in water and dilute to 1000 ml. Filter the solution if it is not
water clear. It is stable when stored in a tightly-stoppered alkali-
resistant glass bottle or polyethylene container.

4. Sodium tungstate, 10%. Dissolve 100 g. of molybdate-free
$Na_2WO_4 \cdot 2H_2O$ in water and dilute to 1000 ml.

5. Sulfuric acid, ⅔ N. Dilute 18.5 ml. of concentrated sulfuric acid,
sp. gr. 1.84, to 1000 ml. with water.

6. Tungstic acid solution. To 800 ml. of water add, with mixing, 50
ml. of 10% sodium tungstate, 0.05 ml. of 85% o-phosphoric acid and
50 ml. of ⅔ N sulfuric acid. This solution is stored in a borosilicate or
polyethylene bottle and is usually stable for months (13). If the
solution becomes cloudy it should be discarded. A fine yellow sediment
may form occasionally on glass surfaces but this does not interfere
with the use of the reagent.

7. Uric acid stock standard, 100 mg. per 100 ml. Transfer 1.000 g. of uric acid to a 1000-ml. volumetric flask. Dissolve separately 0.60 g. of lithium carbonate, Li_2CO_3, in 150 ml. of warm water, filter, and heat the filtrate to 60°C. A slight turbidity at this point is ignored. Add the warm lithium carbonate solution to the flask and mix until the uric acid is completely dissolved. Solution should be complete in 5 minutes; the flask may be warmed under running hot water if necessary, then held under running cold water to cool. The solution usually remains slightly turbid. Add 20 ml. of 40% formaldehyde (formalin). Dilute to about 500 ml. with water, then add slowly and with mixing 25 ml. of 1 N sulfuric acid (prepared by diluting concentrated acid 1 : 36). Dilute to the mark with water, mix well, and store in a brown bottle in the cold (14). This standard is stable up to 5 years or longer.

8. Dilute uric acid working standard, 0.5 mg. per 100 ml. Transfer 0.50 ml. of uric acid stock standard to a 100-ml. volumetric flask and dilute to the mark with water. Store in a refrigerator. Prepare fresh each week.

Procedure for Serum

Add, with mixing, 9.0 ml. of tungstic acid solution to 1.0 ml. of serum. Centrifuge or filter.

NOTE: The rate of addition of the tungstic acid solution is not critical. Equal results are obtained with two other techniques for precipitating proteins: (*a*) Add 1.0 ml. of serum, drop by drop with mixing, to 9.0 ml. of tungstic acid solution. (*b*) To 1.0 ml. of serum add 8.0 ml. of water, 0.5 ml. of 2/3 N sulfuric acid, then 0.5 ml. of 10% sodium tungstate, slowly and with mixing. Precipitations are done in small Erlenmeyer flasks to facilitate mixing. Centrifuging will provide sufficient protein-free solution to permit repeating the analysis for high values.

Transfer 5.0 ml. of clear supernatant or filtrate to a test tube or cuvette. Transfer 5.0 ml. of water to a second tube for a blank and 5.0 ml. of dilute standard (0.5 mg. per 100 ml.) to a third tube. To each tube add 1.0 ml. of 10% sodium carbonate solution and mix. Let stand 10 minutes. Add 1.0 ml. of dilute phosphotungstic acid, mix immediately, and let stand 30 minutes. Measure the absorbance of the standard and unknown within the next 20 minutes against the blank in a spectrophotometer set at 700 mμ. Any photoelectric colorimeter equipped with a red filter may be used (nominal wavelength 640–720 mμ).

Procedure for Urine

If the urine is cloudy, warm a portion of the well-mixed specimen at 60°C. for 10 minutes to dissolve any precipitated urates. Dilute 1.0 ml. of urine to 100 ml. with water. Use 5.0 ml. of the diluted urine and proceed as described for the protein-free filtrate of serum. The absorbance should read between 0.2 and 0.8; otherwise, the urine is diluted appropriately and reanalyzed to obtain readings in this range. The concentration of nonuric acid chromogens in urine is relatively greater than in serum (8); however, the direct colorimetric method is useful in some applications.

Calculations

A standard curve should be prepared to determine the extent of agreement with Beer's law for the particular instrument employed. When observed absorbances fall on the linear portion of the curve, the concentration of uric acid in the unknown may be calculated from the equation:

$$\frac{\text{Absorbance of unknown}}{\text{Absorbance of standard}} \times 5 = \text{mg. of uric acid per 100 ml. serum.}$$

For urine specimens initially diluted 1 : 100,

$$\frac{\text{Absorbance of unknown}}{\text{Absorbance of standard}} \times 50 = \text{mg. of uric acid per 100 ml. urine.}$$

The standard calibration curve usually follows Beer's law to a concentration equivalent to 8 or 10 mg. of uric acid per 100 ml. of serum. Higher values may be taken from the curve or the analysis may be repeated starting with 2.0 ml. of filtrate + 3.0 ml. of water. In the latter procedure, final results are multiplied by 5/2.

A standard calibration curve is prepared as follows: Transfer 3.00 ml. of uric acid stock standard (100 mg. per 100 ml.) to a 200-ml. volumetric flask and dilute to the mark with water to provide a dilute standard containing 1.5 mg. per 100 ml. Prepare working standards by diluting 1.0, 2.0, 3.0, 4.0, and 5.0 ml., respectively, of the dilute standard to 5.0 ml. of water. Proceed with color development of the standards as described under procedure, including a blank. These standards correspond to 3, 6, 9, 12, and 15 mg. of uric acid per 100 ml. of serum in the method presented. On linear graph paper plot the absorbance for each standard against the equivalent concentration of uric acid expressed as mg. per 100 ml. of serum. If desired, all readings may be taken in % transmittance. In this event the blank is set to 100% transmittance and the transmittance readings for the standards are plotted on semilog paper.

Discussion

Recovery of uric acid added to serum prior to precipitation of protein ranged from 80–95% in agreement with the findings of others (8, 15, 16, 17). The % recovered is decreased as the amount added is increased. Some of the added uric acid is precipitated with the protein and may be recovered by dissolving the precipitate in 1 ml. of 0.1 N sodium hydroxide, diluting to 9 ml. with water, and reprecipitating the protein by adding 1 ml. of 0.1 N hydrochloric acid. The filtrate is analyzed in the usual manner with allowance made for entrapped solution in the precipitate from the first precipitation. When this technique was applied to serum to which had been added the equivalent of 1 to 10 mg. of uric acid per 100 ml. of serum, recoveries averaged 100% (range 98–102%). This same technique was applied to twenty-six serums to which no uric acid had been added. The concentrations of uric acid ranged from 1.2–15.9 mg. per 100 ml. The additional uric acid recovered by the second precipitation ranged from —0.05–0.26 mg. per 100 ml. (mean: 0.10; s ± 0.09). The % of total uric acid recovered by the first precipitation ranged from 95.5–101% (mean 98.0; s ± 1.5). These results indicate that uric acid is recovered adequately from serum by the single precipitation technique. It is not clear why the % recovery of uric acid added to serum *in vitro* is so much less than the recovery of even high concentrations present in nontreated serum.

To determine the precision of the method, twenty separate aliquots from a quantity of pooled serum were analyzed. The results obtained varied from 5.59 to 5.78 mg. per 100 ml. (mean 5.69; s ± 0.06). The absorbance of the blank is usually less than 0.005 in this method.

To assess the accuracy of the method a comparison was made with the cyanide method of Brown (18) as described in Standard Methods of Clinical Chemistry, Vol. I (15). Serum (fifty specimens) was deproteinized as described above and the same protein-free filtrates were analyzed by both methods. Values obtained by the cyanide procedure ranged from 1.16–11.74 mg. per 100 ml. (mean 5.18). Values obtained by the carbonate procedure ranged from 1.18–13.24 mg. per 100 ml. (mean 5.54). Regression analysis of y (mg. per 100 ml. by carbonate method) on x (mg. per 100 ml. by cyanide method) yielded the equation: $y = 0.97x + 0.52$. Results obtained by the carbonate procedure averaged 7% higher than those obtained by the cyanide procedure.

The method was also compared with a uricase procedure (19) on twenty-four serums. Values obtained by the uricase procedure ranged from 2.18–10.10 mg. per 100 ml. (mean 5.51). Values obtained by the carbonate procedure ranged from 2.67–10.50 mg. per 100 ml. (mean 5.98). Regression analysis of y (mg. per 100 ml. by carbonate method) on x (mg. per 100 ml. by uricase method) yielded the equation: $y = 0.95x + 0.75$. Results obtained by the carbonate procedure averaged 8% higher than those obtained by the uricase procedure.

Johnstone (20) concluded that results obtained with Brown's method agree within 0.2 mg. per 100 ml. of results obtained by a specific uricase method for measuring uric acid. The data presented above are in substantial agreement with this conclusion. Inasmuch as the incorporation of cyanide into the method increases the absorbance per mg. of uric acid by several fold, it is apparent that the relative effect of interfering substances would be considerably less in the methods employing cyanide.

Ascorbic acid is one major interfering substance found in fresh serum. This interference is not significant in the more sensitive cyanide methods but is of importance in methods where cyanide is replaced by sodium silicate or carbonate (21). Incubation for 10 minutes after adding sodium carbonate effectively destroys ascorbic acid and catechols present in serum and urine. When this step is omitted, the apparent uric acid concentration will be elevated by 1 mg. per 100 ml. for each mg. of reduced ascorbic acid present.

Full color development in the procedure described depends on the presence of sufficient carbonate to bring the pH of the final reaction mixture to 10.0–10.4. This requires a balance between the sodium carbonate and the acidity contributed by the protein-free filtrate and the phosphotungstic acid. An increase in the concentration of sodium carbonate results in more rapid color development but the period of constant absorbance is decreased. An increase in the concentration of both phosphotungstic acid and carbonate will result in better agreement with Beer's law at higher concentrations but turbidities tend to develop. This has been prevented in one modification by incorporating lithium sulfate in the phosphotungstic acid reagent (8).

Temperature control is not critical. Final reaction mixtures were incubated for 30 minutes at 20, 25, and 30°C., respectively, with equal results. The final absorbance obtained was also the same after 30 minutes for mixtures incubated in the dark or exposed to sunlight.

Normal Values

Normal values reported for uric acid in serum vary considerably (15). Representative values are listed in Table I together with values obtained by the present method in the submitter's laboratory on ninety-six apparently normal adults. A reasonable set of normal values for uric acid in serum would appear to be: males, 3–7 mg. per 100 ml.; females, 2–6 mg. per 100 ml. The daily variation in the serum uric acid of an individual is approximately 0.4–0.5 mg. per 100 ml. (22). Normal serum levels are slightly lower on subjects receiving a low purine diet (23). The clinical significance of uric acid has been reviewed elsewhere (24, 25).

TABLE I

Normal Levels of Uric Acid in Human Serum or Plasma

Author	Method	Sex	Number	Mean (mg. per 100 ml.)	Range (mg. per 100 ml.)
Gjørup et al. (26)	Uricase	M	143	5.04	2.6–7.5
		F	157	3.84	2.0–5.7
Fechtmeier and Wrenn (27)	Uricase	M	38	5.89	4.2–7.6
		F	40	3.75	1.5–6.0
Johnstone (20)	Uricase	M	17	4.61	2.8–7.0
		F	21	3.75	2.0–6.5
Yü and Gutman (28)	Uricase	M	49	5.4	3.0–7.8
		F	53	4.1	2.3–5.9
Bensley et al. (16)	Cyanide	M	63	4.8	2.9–6.9
		F	47	4.0	2.9–5.7
Wolfson et al. (29)	Cyanide	M	22	5.27	3.6–6.5
		F	20	4.05	2.6–5.5
Alper and Seitchik (21)	Silicate (7)	M	95	5.4	3.8–7.1
		F	48	4.0	2.6–5.4
Hendry (30)	Carbonate (6)	M	300	5.4	4–7
		F	100	4.3	3–6
Caraway (present study)	Carbonate	M	49	5.0	3.1–7.3
		F	47	4.1	1.9–6.3

REFERENCES

1. Jorgensen, S., and Theil Nielson, AA. Uric acid in human blood corpuscles. *Scand. J. Clin. Lab. Invest.* **8,** 108–112 (1956).
2. Folin, O., and Denis, W., A new (colorimetric) method for the determination of uric acid in blood. *J. Biol. Chem.* **13,** 469–475 (1912–13).
3. Benedict, S. R., The determination of uric acid in blood. *J. Biol. Chem.* **51,** 187–207 (1922).
4. Heilmeyer, L., and Krebs, W., Bestimmung der Harnsäure im Blutserum mit dem Zeissschen Stufenphotometer unter besonderer Berücksichtigung der optischen Grundlagen. *Biochem. Z.* **223,** 365–372 (1930).
5. Kern, A., and Stransky, E., Beitrag zur kolorimetrischen Bestimmung der Harnsäure. *Biochem. Z.* **290,** 419–427 (1937).
6. Caraway, W. T., Determination of uric acid in serum by a carbonate method. *Am. J. Clin. Pathol.* **25,** 840–845 (1955).
7. Archibald, R. M., Colorimetric measurement of uric acid. *Clin. Chem.* **3,** 102–105 (1957)
8. Henry, R. J., Sobel, C., and Kim, J., A modified carbonate-phosphotungstate method for the determination of uric acid and comparison with the spectrophotometric uricase method. *Am. J. Clin. Pathol.* **28,** 152–160; corr. 645 (1957).
9. Hausman, E. R., Lewis, G. T., and McAnally, J. S., Determination of uric acid. A note. *Clin. Chem.* **3,** 657–658 (1957).
10. Lous, P., and Sylvest, O., A comparison of three methods, utilizing different principles, for the determination of uric acid in biological fluids. *Scand. J. Clin. Lab. Invest.* **6,** 40–42 (1954).
11. Kanabrocki, E. L., Greco, J., Wikoff, L., and Veach, R., Comparison of plasma uric acid levels obtained with five different methods. *Clin. Chem.* **3,** 156–159 (1957).
12. Caraway, W. T., Chlorine in distilled water as a source of laboratory error. *Clin. Chem.* **4,** 513–518 (1958).
13. Caraway, W. T., Stabilized tungstic acid reagent for blood deproteinization. *Chemist-Analyst* **47,** 44–45 (1958).
14. Vincent, D., and Mignon, M., A propos du dosage de l'acide urique dans le sang. *Ann. Biol. Clin. (Paris)* **15,** 510–513 (1957).
15. Natelson, S., and Kaser, M., Uric acid. *In* "Standard Methods of Clinical Chemistry," (M. Reiner, ed.), Vol. I, pp. 123–135. Academic Press, New York, 1953.
16. Bensley, E. H., Mitchell, S., and Wood, P., Estimation of uric acid in serum and whole blood by an electrophotometric modification of Folin's method. *J. Lab. Clin. Med.* **32,** 1382–1386 (1947).
17. Steel, A. E., The determination of uric acid in biological materials. *Biochem. J.* **68,** 306–309 (1958).
18. Brown, H., The determination of uric acid in human blood. *J. Biol. Chem.* **158,** 601–608 (1945).
19. "Descriptive Manual No. 10," Worthington Biochemical Corp., Freehold, New Jersey, 1958.
20. Johnstone, J. M., True uric acid values. *J. Clin. Pathol.* **5,** 317–318 (1952).

21. Alper, C., and Seitchik, J., Comparison of the Archibald-Kern and Stransky colorimetric procedure and the Praetorius enzymatic procedure for the determination of uric acid. *Clin. Chem.* **3**, 95–101 (1957).

22. Zauchau-Christiansen, B., The variation in serum uric acid during 24 hours and from day to day. *Scand. J. Clin. Lab. Invest.* **9**, 244–248 (1957).

23. Crone, C., and Lassen, U. V., Some uric acid values in normal human subjects. *Scand. J. Clin. Lab. Invest.* **8**, 51–54 (1956).

24. Bishop, C., and Talbott, J. H., Uric acid: Its role in biological processes and the influence upon it of physiological, pathological and pharmacological agents. *Pharmacol. Rev.* **5**, 231–273 (1953).

25. Gutman, A. B., and Yü, T. F., Gout, a derangement of purine metabolism. *Advan. Internal Med.* **5**, 227–302 (1952).

26. Gjørup, S., Poulson, H., and Praetorius, E., The uric acid concentration in serum determined by enzymatic spectrophotometry. *Scand. J. Clin. Lab. Invest.* **7**, 201–203 (1955).

27. Fechtmeier, T. V., and Wrenn, H. T., Direct determination of uric acid using uricase. *Am. J. Clin. Pathol.* **25**, 833–839 (1955).

28. Yü, T. F., and Gutman, A. B., Quantitative analysis of uric acid in blood and urine. Methods and interpretations. *Bull. Rheumatic Diseases* **7**, (Suppl.) S17–S20 (1957).

29. Wolfson, W. Q., Hunt, H. D., Levine, R., Guterman, H. S., Cohn, C., Rosenberg, E. F., Huddlestun, B., and Kadota, K., The transport and excretion of uric acid in man. V. A sex difference in urate metabolism. *J. Clin. Endocrinol.* **9**, 749–767 (1949).

30. Hendry, P. I. A., White, K. H., and Stanger, I. J., Serum uric acid. *Med. J. Australia* **II**, 956–959 (1959).

AUTHOR INDEX

Numbers in parentheses are reference numbers and are included to assist in locating references when authors' names are not mentioned in the text. Numbers in italics refer to pages on which the references are listed.

A

Abdulnabi, M., 11(13), *13*
Ackermann, P. G., 93, *98*
Acland, J. D., 135, 137, *138*
Adams, M., 19(6), *21*
Adelstein, S. J., 45(17), *46*
Albanese, A. A., 11, *13*
Albright, F., 159(35), *162*
Albrink, M. J., 95, 96, *98*, *99*
Alfheim, A., 72(9), 79(9), 83(18), *82*, *83*
Allen, W. M., 74(11), *82*, 159(37), *162*
Alles, G. A., 47(3), *55*
Alonzo, N., 87(24), 90, *98*
Alper, C., 244(21), 245, *247*
Amelung, D., 163(13), *171*
Ammon, R., 47(4), 49(4), *55*
Andres, E., 92, 93, *98*
Anger, V., 86(7), *97*
Anton, H. U., 157(24), 158(24), 159(24), *161*
Antunes, L. N., 158(30), *161*
Appleby, J. K., 118(16), *120*
Apt, L., 139, 141(2), 143(2), 144(2), 145(28), 148(28), *149*, *150*
Archibald, R. M., 8, *13*, 239(7), 245(7), *246*
Ascoli, I., 173(6), 174(6), 175(6), *181*
Ashenbrucker, H., 143(7), *149*
Augustinsson, K. B., 47(5), 49(5), *55*
Aylward, F. X., 95(34), *98*

B

Babbott, D., 54(17), *56*
Bakwin, H., 62(10), *63*
Bakwin, R. M., 62(10), *63*

Bang, I., 173, *181*
Barkan, G., 139, 144(1,3), *149*
Barker, S. B., 126, 137, *137*
Barton, D. H. R., 114(11), *120*, 155(12), *161*
Batt, W. G., 101(3), *111*
Bauer, F. C., Jr., 85(11), 86(9,11), 91(11), 93(11), *97*
Bauld, W. S., 65(5), 66(7), 75(7), 76, 81(16), *82*
Bayer, E., 86, *98*
Beach, E. E., 101, 107, 111(11), *112*
Bean, W. B., 159(33), *162*
Bearn, A. G., 171(26), *172*
Beher, W. T., 159(36), *162*
Bellucci, A., 52(16), 53(16), 54(16), *56*
Benedict, S. R., 239(3), *246*
Benenson, A. S., 228, *234*
Bensley, E. H., 243(16), 245, *246*
Bentley, R., 101(5), *111*
Berkman, S., 163(20), 167(20), 169(20), *172*, 223(21), *234*
Berkowitz, J., 23(1), 29(6), *29*
Berson, B., 144(16), *150*
Bessey, O., 144(16), *150*
Bethard, W., 148(33), *150*
Beveridge, J. M. R., 191, *195*
Beyers, M., 148(31), *150*
Bishop, C., 245(24), *247*
Bitman, J., 155(13), *161*
Black, P., 148(30), *150*
Blahey, P. R., 83(18), *83*
Blair, H. A. F., 72(10), 79(10), *82*
Blix, G., 96, *99*
Bloor, W. R., 85(5), *97*, 173, *181*
Bobansky, O., 163(7), *171*

Pletscher, A., 198(10), 200(10), *203*
Plotner, K., 144(13), *149*
Polonovski, J., 86, *97*
Pontius, D., 157(24), 158(24), 159(24), *161*
Popper, H., 47(1), 51(1), *55*
Porter, C. C., 113, 114(2,4), 118(4), *119*
Poulson, H., 245(26), *247*
Power, M. H., 192(7), *195*
Praetorius, E., 245(26), *247*
Pratt, E. L., 160, *162*
Pricer, W. E., Jr., 86, *97*

R

Raabo, E., 102, 103(18), *112*
Raeside, J. I., 83(18), *83*
Ramasarma, G. B., 12(24), *13*
Ramsay, W., 144(19,21), *150*
Rappaport, F., 157(28), 158(28), 159(28), *161*
Rapport, M. M., 87(24), 90, *98*, 197(1), *202*
Rath, C., 148, *150*
Rausch, V. L., 11(14), *13*
Ravin, H. A., 44(15), *46*
Reddy, W. J., 113, 118(8,14), *120*
Reifeinstein, E., 159(35), *162*
Reinfrank, R., 52(15), *56*
Reinhart, H. L., 191(3), *195*
Reinhart, R., 144(24), *150*
Reinhold, J. G., 15(3), 16, *21*, 47(9), *55*
Reinwein, H., 23(3), *29*
Ressler, N., 144(25), *150*
Reuther, K. H., 86, *98*
Rice, E. W., 39, 40(3,6), 41(3), 42(3,6), 44(3), 45(3,18,19), *45, 46*, 57(3), 62, 63(13,14), *63*, 102, 103, *112*
Richman, A., 96, *99*
Rieben, W. K., 175(11), *182*
Riggs, D. S., 125, 126, 130, 136(3), 137(3), *137*
Ritter, J. J., 43(13), *46*
Roberts, E., 12(24), *13*
Rogers, J., 155(16), *161*
Ronzoni, E., 12(24), *13*

Rosenberg, E. F., 245(29), *247*
Rosenthal, H. L., 86, 88, *97*
Ross, C., 171(28), *172*
Ross, J. R., 139, 141(2), 143(2), 144(2), 145(28), 148(28), *149, 150*
Routh, J. I., 191(8), 192(8), 194(8), *195*
Russell, J. A., 1(1), 11(1), *12*, 102(20), 103(20), 105(20), 107(20), 110(20), *112*
Ryan, J., 155(11), 157(11), *160*

S

Saifer, A., 101, 102, 107, *111*
Salomon, L. L., 101, 102(9), 103(9), 111(9), *111*
Salter, W. T., 126, *137*, 157, 158(23), 159(23), *161*
Samuels, L. T., 113, *120*
Sandell, E. B., 125, *137*
Sato, T., 42(9), *46*
Schade, A., 143(11), 144(24), 148(11), *149, 150*
Schaffert, R. R., 137, *138*
Schales, O., 139, 143(4), 144(4), *149*
Schapira, F., 143(10), *149*
Schapira, G., 143(10), *149*
Schedl, H. P., 159(33), *162*
Scheidt, U., 23(3), *29*
Scheinberg, I. H., 39(1,2), 40(1), 45(1,16), *45, 46*, 62, 63(12), *63*
Schenker, S., 170(24), *172*
Schiff, L., 170(24), *172*
Schimizu, M., 42(9), *46*
Schlenk, F., 163, *171*
Schmitz, G. H., 43(13), *46*
Schneider, A. J., 210(7), 219(7), *234*
Schneider, I., 157(22), 159(22), *161*
Schön, H., 86, 87, 92, *97*
Scholler, J., 163(7), *171*
Schubert, W. K., 62(9), *63*
Schuette, H., 144(22), *150*
Schumacher, E. R., 159(33), *162*
Schwert, R. S., 86(21), *98*
Scism, G. R., 158(29), *161*
Seacy, R., 157(27), 158(27), 159(27), *161*

Seckfort, H., 92, 93, *98*
Segalove, M., 1(2), 12(2), *12*, 42(7), 44(7), *46*, 223(18,21), 228, *234*
Segar, A. J., 20(7), *21*
Segovia, R., 114(11), *120*
Seitchik, J., 244(21), 245, *247*
Seligson, D., 104, 106, *112*
Shapiro, B., 85, 86, 88, *97*
Shinowara, G. Y., 191, *195*
Shore, E., 191, *195*
Silber, R. H., 113(6), 114(2,4), 118(4), 119(3), *119*, *120*
Sister Lenta, M. Petra, 163(5), *171*
Sister Riehl, Agatha, 163(5), *171*
Sjoedsam, A., 198(12), *203*
Sjoerdsma, A., 121(2,3,4,6), 123(2), *124*
Skanse, B., 137, *138*
Skupp, S. J., 85(6), 96, *97*
Slade, C. I., 131, *137*
Slaunwhite, W. R., 159(32), *162*
Sloan, C. H., 158(31), *162*
Sloan-Stanley, G. H., 173(7,8), 174(7,8), 176(7,8), 179(8), 180(8), *182*
Smyles, E., 155(3), *160*
Smyrniotis, F., 170, *172*
Snedecor, G. W., 219(15), *234*
Sobel, C., 1(2), 12(2), *12*, 239(8), 242(8), 243(8), 244(8), *246*
Somogyi, M., 103, 105, 108, 110, *112*, 129, *137*
Sowa, M., 225(22), *234*
Spear, F. E., 102(17), 103(17), *112*
Sperry, W. M., 88, *99*, 175(9,10), 176(12), 177(9), 179(13), 180(9,13), *182*
Spoerri, P. E., 87, 91, *98*
Sprague, A., 160(40), *162*
Sprague, A. L., 37(7), *38*
Stanger, I. J., 245(30), *247*
Stark, G. R., 57, 60(2), *63*
Steel, A. E., 243(17), *246*
Stein, W. H., 12(23), *13*
Stengle, J., 143(11), 148(11), *149*
Stern, I., 85, 86, 88, *97*
Stern, S., 158(29), *161*

Sternlieb, I., 39(1), 40(1), 45(1,16), *45*, *46*, 62, 63(12), *63*
Stewart, C. P., 85(3), *97*
Stevenson, B. M., 159(33), *162*
Stimson, M., 23(2), *29*
Stoddard, J. D., 85(4), *97*
Stransky, E., 239(5), *246*
Strong, J. A., 81(15), *82*
Stubbs, R. D., 118(15,16), *120*
SubbaRow, Y., 191, *195*
Sulkowitch, H., 159(35), *162*
Summers, R. M., 62, *63*
Sunde, C. J., 42, *46*
Suntzeff, V., 163(6), *171*
Sutherland, E. W., 201, *203*
Sylvest, O., 239, *246*

T

Talbot, N. B., 155(11), 157(11), *160*
Talbott, J. H., 245(24), *247*
Tammes, A. R., 191(8), 192(8), 194(8), *195*
Taussky, H. H., 191, *195*
Teich, S., 155(16), *161*
Teller, J. D., 102, *112*
Tennant, R., 51(10), 52(10,14,16), 53(16), 54(16), *55*, *56*
Tennant, W. S., 91(29), 96, *98*
Terkildsen, T. C., 102, 103(18), *112*
Terry, L. L., 121(3,4), *124*
Thiers, R. E., 165(18), 167(19), *172*
Thomas, P. Z., 160(39), *162*
Thompson, L. K., 228(24), *234*
Thompson, R. C., 11(13), *13*
Thomson, W. S. T., 11, *13*
Thorn, G. W., 113(8), 118(8), *120*
Thuline, H. C., 157(25), 159(25), *161*
Timbres, H. Q., 95(34), *98*
Tinguely, R., 148(34), *150*
Titus, E., 121(7), 122(7), *124*, 197(2), *202*
Toh, C. C., 197(4), *202*
Tompsett, S. L., 91(29), 96, *98*
Toralballa, G. C., 19(6), *21*
Toro, J., 93(32), *98*
Totterman, L. E., 157(19), *161*
Tourigny, L. G., 47(9), *55*

258

Yü, T. F., 245(25), *247*
Yushok, W. D., 101(3), 102, *111, 112*

Z

Zak, B., 144(25), *150*

Zauchau-Christiansen, B., 245(22), *247*
Zilversmit, D. B., 90, 91, 96, *98*
Zimmerman, H. J., 163(11,14), *171*
Zimmermann, W., 151, 157(24), 158(24), 159(24), *160, 161*
Zondek, B., 83(20), *83*

SUBJECT INDEX

A

Acetylcholinesterase, 51
Adrenocorticotropic hormone, 12
Amino acids, plasma, 1-8
  plasma values, 11, 12
  urine, 1-3, 9-10
Aminoaciduria, 12
Amylase, normal values in serum and
  urine, 21
  stability in serum, 19
  turbiclimetric measurement, 15-22
Androsterone, 151, 155
Arginine, 12
Arteriosclerotic disease, 95
Aspartic acid, 1

B

Bladder cancer, 170
Blood, *see also* plasma and serum
  carbon monoxide determination,
    31-38
  oxygen saturation, 183-190
  uric acid, 239
Blood clotting, 96

C

Calcium oxalate, infrared spectra, 25,
  27, 28
Calculus, identification by spectroscopy,
  23-30
Carbonato-apatite, infrared spectra, 25,
  27, 28
Carbon monoxide, spectrophotometric
  determination of in blood, 31-38
Carboxyhemoglobin, 183
  extinction coefficient, 32-34
  spectral absorbance, 31-32
Carcinomas, 170
  cholinesterase, 53, 54-55
Ceruloplasmin, assay, 39-46
  normal values, 44
  pathological values, 45

Cholesterol, 91, 94, 95
Cholinesterase, 47-56
  normal values, 52
  pathological values, 53
Cirrhosis, 54, 63
Clinical chemistry, statistics applied,
  205-238
Copper, assay in serum, 57-63
  plasma, 39
  range of values, 62-63
Corticosteroid, normal values, 119
Cortisone, 113
Creatinine, 119
Cystine, 12
  infrared spectra, 27
Cystinuria, 12

D

Dehydroepiandrosterone, 155
Dehydroisoandrosterone, 151, 155
de Toni-Fanconi sydrome, 12
Diabetes, 95
Diabetes mellitus, 12
17, 21-Dihydroxy-20-ketosteroid, 118

E

Ergothioneine, 293
Estriol, 65
  pregnancy urine, 74-79
Estrogens, nonpregnancy urine, 67-72
  normal values, 80-81
  pregnancy urine, 72-74
  urinary, 65-84
Etiocholanolone, 151, 155

F

Fatty acid, esterified in serum, 85-100

G

Galactosemia, 12
Gastric cancer, 170
Glucose, enzymatic assay, 101-112
  normal values in blood, 107-111

Plasma amino nitrogen, normal value, 11
pathological value, 12
Pregnancy, estrogens, 72-74, 80-81
serum copper, 62
Proline, 1
Protein, precipitation 126
Pulmonary emboli, 170

R
Reagent blanks, 228-229
Rheumatic heart disease, 63
Rheumatoid arthritis, 63

S
Scleroderma of small bowel, 45
Serotonin, 121
content of various tissues, 202
estimation of, 197-204
Serum, amylase determination, 15
amylase stability, 19
ceruloplasmin assay, 39-46
cholinesterase, 47-56
copper, 57-63
esterified fatty acid, 85-100
glucose, 101-112
inorganic phosphorus, 191-196
iron, 139-144
iron-binding capacity, 144-149
lactic dehydrogenase, 163-172
protein-bound iodine, 125-138
total lipids, 173-182
triglyceride, 85-100
uric acid, 241
Sphingomyelin, 90
Sprue, 45
Standards, 224-228
Standard deviation, 206, 207-209

Statistics, analysis of method, 215-223
application to clinical chemistry, 205-238
frequency distributions, 205-206
miscellaneous tests, 229-233
normal range, 209-215
standard deviation, 207-209

T
Thyroxine, 135
Triglyceride, serum, 85-100

U
Urea, 1, 174
removal with urease, 8, 9
Uric acid, 239-248
infrared spectra, 26
normal values, 245
Urinary amino nitrogen, normal values, 11-12
pathological values, 12
Urinary tract calculi, identification of by infrared spectroscopy, 23-30
Urine, amino acid determination, 1-14
amylase determination, 15
estrogens, 65-84
glucose assay, 111
17-hydroxycorticosteroids, 113-120
5-hydroxyindoleacetic acid, 121-124
17-ketosteroids, 151-162
uric acid, 242

V
Van Slyke-Neill method, 1-4
Viral heptatitis, 54

W
Wilson's disease, 12, 45, 63